Vanguard of Valor

Small Unit Actions in Afghanistan

General Editor
Donald P. Wright, Ph. D.

Afghan Study Team

Anthony E. Carlson, Ph. D.

Michael J. Doidge

Scott J. Gaitley

Kevin M. Hymel

Matt M. Matthews

John J. McGrath

Lieutenant Colonel John C. Mountcastle

Ryan D. Wadle, Ph. D.

Library of Congress Cataloging-in-Publication Data

Vanguard of Valor: Small Unit Actions in Afghanistan / General Editor, Donald P. Wright; Afghan Study Team, [Et Al.].

P. Cm.

1. Afghan War, 2001---Campaigns. 2. United States. Army--History--Afghan war, 2001- 3. United States. Army--Civic Action. 4. United States--Armed Forces--Stability Operations. 5. Afghanistan--History, Military--21st Century. Wright, Donald P., 1964- II. Title: Small Unit Actions in Afghanistan.

DS371.412.V36 2012

958.104'7420973--dc23

2011052370

Second printing

Foreword

Since 2001, the US Army in Afghanistan has been conducting complex operations in a difficult, often dangerous environment. Living in isolated outposts and working under austere conditions, US Soldiers have carried out missions that require in equal parts a warrior's courage and a diplomat's restraint. In the larger discussions of the Afghanistan campaign, the experiences of these Soldiers—especially the young sergeants and lieutenants that lead small units—often go undocumented. But, as we all know, success in Afghanistan ultimately depends on these small units and their leaders, making their stories all the more important.

In 2010, as the scale and tempo of Coalition operations in Afghanistan increased, so did the need for historical accounts of small-unit actions. As commander of the International Security Assistance Force (ISAF), I commissioned the Combat Studies Institute to research and write the cases collected in this volume and in those that will follow. By capturing key insights from both lethal and non-lethal operations, I hoped these accounts would be of immediate utility to sergeants and lieutenants at the center of future operations.

The eight actions described in these pages take the reader through a wide range of platoon-level operations, from an intense firefight near Kandahar to an intricate civic action project in Kunar Province. Drawing from dozens of Soldier interviews, these accounts vividly depict the actions themselves and offer critical insights of greatest benefit to the small-unit leaders of today and tomorrow.

The US Army always has prided itself as an institution of constant learning, strongly committed to drawing lessons from its past. This volume from the Combat Studies Institute is an excellent example of that long and honorable tradition. I hope that you will find the actions related in Vanguard of Valor to be both instructive and compelling. I am sure that you will find them to be inspirational.

David H. Petraeus
General, United States Army (Retired)
Director, Central Intelligence Agency

Acknowledgements

In late 2010, the Commander of the International Security Assistance Force (ISAF), General David Petraeus, asked the Combat Studies Institute (CSI) to publish a series of historical accounts focused on tactical-level operations in Afghanistan. For General Petraeus, the use of history in the course of an ongoing campaign had immediate utility. His goal was to make recent operations in Afghanistan relevant to our junior leaders who were soon to conduct similar operations, often on the same terrain. This volume is the first in a series that should do just that.

In making General Petraeus' vision a reality CSI had a great deal of assistance. US Forces-Afghanistan (USFOR-A) provided the initial funding for the establishment of the Afghan Study Team at CSI and later the US Army Center of Military History (CMH) provided funding to continue this important work. The USFOR-A Staff offered guidance and coordinated the research effort that allowed for the collection of data. Historians from CMH, as well as those in several Military History Detachments (MHDs) that operated in Afghanistan, gathered the lion's share of the documents we used. I extend a special thanks to the numerous Soldiers who sat for interviews thus enabling us to weave first-hand accounts of the various actions into the studies.

Finally, I want to acknowledge the writers and editors that comprise the Afghan Study Team. As a group, they quickly established techniques and norms that allowed them to write contemporary military history that is lively, relevant, and meets the high expectations of the historical profession. These studies will be of great use to our Army's officers and NCOs who are currently conducting operations around the world; and, they will well serve tomorrow's leaders who will likely face similar challenges in conflicts that are yet to appear on the horizon.

CSI – The Past is Prologue!

Roderick M. Cox
Colonel, US Army
Director, Combat Studies Institute

Contents

List of Figures

Chapter 6.

Chapter 7.

Chapter 8.

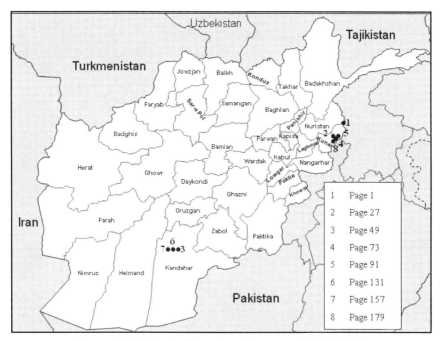

Geographical Key to Operations Recounted in this Work.

Firefight above Gowardesh
by
Lieutenant Colonel John C. Mountcastle

In the early evening hours of 21 June 2006, 16 American Soldiers from the 3d Squadron, 71st Cavalry (3-71 CAV) of the US Army's 10th Mountain Division engaged in a desperate fight against enemy insurgents high on a mountain ridge overlooking the Gremen Valley in the northeast corner of Afghanistan. After 30 minutes of intense combat, two American Soldiers were dead and one more was critically wounded. The enemy element that attacked the US force withdrew after suffering significant casualties. What happened on this ridgeline high above the small village of Gowardesh that evening demonstrated the daunting tactical and operational challenges facing the US Army in the mountainous terrain along the border between Afghanistan and Pakistan. These challenges were persistent and would define operations in the region for years to come. The engagement also demonstrated the capability of Coalition forces to meet and defeat a well-equipped, trained, and determined enemy on his home terrain.

Three years after the firefight, one of the leaders of the American detachment, Sergeant First Class Jared C. Monti, was posthumously awarded the Medal of Honor for his actions in the battle but the heroism and valor he demonstrated during the engagement were showcased by all of the Soldiers that fought for one another that evening. In order to gain a better appreciation for what they endured, this study offers the details of the events leading up to, during, and after the engagement. It also examines a number of important operational and tactical decisions that shaped the events leading up to the insurgent attack on the American "*kill team*" from 3-71 CAV. In doing so, this study illustrates the complex and challenging nature of conducting counterinsurgency in Afghanistan, and in particular, the mountainous region of Nuristan and Kunar Provinces that serves as the primary corridor to Pakistan for Taliban and other foreign fighters. These are the types of tactical and operational challenges American Soldiers can well expect to encounter in the near future.

Background

In April of 2006, Combined Joint Task Force *76* (CJTF-76), under the leadership of the 10th Mountain Division, initiated Operation MOUNTAIN LION in eastern Afghanistan aimed at defeating insurgent operations and establishing Coalition control in the region. A key aspect of the operation was the utilization of a "clear-hold-build-engage" approach which a former

G3 of the 10th Mountain Division referred to as a "score with four essential parts."[1] As a part of a counterinsurgency effort, the "clear" step focused on using lethal military operations to clear targeted areas of any enemy presence or control. The next step of "hold" entailed creating a lasting security by establishing Coalition presence in the cleared area. The last two parts of the approach of "build" and "engage" centered on improving infrastructure and facilities for local communities and establishing a working relationship with the people based on communication and a shared trust. With Operation MOUNTAIN LION, CJTF-76 planners hoped to defeat the resurgent Taliban and other enemies of the fledgling Government of Afghanistan that had recently increased their attacks in the provinces of Kunar and Nuristan. In the months that followed, CJTF-76 forces, which included the 3rd Infantry Brigade Combat Team from the 10th Mountain Division, pushed into a number of suspected enemy strongholds establishing contact with the local civilians and building combat outposts to sustain the Coalition's gains. The most notable of these was the establishment of the Korengal Outpost in the heavily-contested Korengal Valley in early May. Fighting with the insurgents was intermittent as most of the enemy chose to avoid the clearing operations and wait for the Coalition forces to return to their bases.

By June 2006, the majority of effort for what remained of Operation MOUNTAIN LION was focused on the "hold" portion of the operation with the transition to "build" and "engage" just beginning. In the latter phases, the Provincial Reconstruction Teams (PRTs), organizations composed of both military and civilian officials dedicated to assisting the local population with building new infrastructure and establishing a new government, would take the lead.[2] However, the security situation in northeast Afghanistan was still unstable. To facilitate the transition to the next phases, the 3d Squadron, 71st Cavalry Regiment (3-71 CAV) of the 3d BCT, 10th Mountain Division operating out of the newly established Forward Operating Base (FOB) Naray, established new checkpoints and outposts in the northeastern tip of Afghanistan once MOUNTAIN LION officially ended on 15 May 2006. FOB Naray was the northernmost US outpost in Afghanistan at the time and, according to one account, Naray District was "a clutch of mud-brick and stone villages inhabited by 30,000 *Pashtun* tribes people" who had not seen outsiders since the Soviet Army departed the region in 1989.[3] In 2006, 3-71 CAV became known as Task Force (TF) *Titan* after the squadron added key attachments to its organic force of four company-sized maneuver elements. The task force's mission was to secure the major population centers in Nuristan and Kunar by clearing areas suspected of harboring insurgents and targeting insurgent

Figure 1 - Northeast Afghanistan.

infiltration routes into the region. In June 2006, CJTF-76 planned to launch a new operation named MOUNTAIN THRUST which aimed to eliminate remaining centers of enemy activity across northeast Afghanistan. The CJTF assigned TF *Titan* the mission of separating what was left of the enemy in southeastern Nuristan Province from the local population and preventing further enemy infiltration into the region from Pakistan. The task force operation, known as GOWARDESH THRUST, became CJTF-76's main effort.

For Operation GOWARDESH THRUST, TF *Titan* planned to execute an operation to clear the Gremen Valley (also known as the Gowardesh Valley). GOWARDESH THRUST would feature two company-sized units moving into the valley using both ground maneuver and air assault. Before launching the operation however, the task force needed to establish surveillance on target areas and named areas of interest (NAIs), locations in the valley where the TF *Titan* staff believed activity would reveal insurgent organization and intent. During OPERATION MOUNTAIN LION, the squadron leadership developed a new and somewhat unorthodox platoon-sized unit just for this type of reconnaissance and observation mission. The squadron combined a Combat Observation and Lasing Team (COLT) section with its sniper section to create a 16-man formation that would have enhanced observation, fire direction, and security capabilities. This unique formation, which became known to the Soldiers of 3-71 CAV as a *kill team*, was described by the squadron operations officer (S3) as a tactically flexible unit that could "maximize the capabilities" of the squadron and defend itself against enemy attack. The *kill team* was comprised of 16 hand-picked Soldiers and NCOs who were, in the S3's words, "the most skilled…the best of the organization."[4] The *kill team* for GOWARDESH THRUST was placed under the leadership of two noncommissioned officers, Staff Sergeant Christopher Cunningham, who oversaw the sniper section, and Staff Sergeant Jared Monti, a section leader from the squadron's COLT platoon. Cunningham and Monti were charged with establishing an observation post (OP) close to the top of Mountain 2610, which made up the western wall of the Gremen Valley. From this position, they were to observe a number of NAIs and to assist in directing artillery fires and close air support for the task force as it conducted its move into the valley. In addition, the *kill team* was supposed to watch for "squirters," a term for insurgents attempting to flee the valley, and engage them as targets of opportunity.[5] In all, the mission was not overly complex. It was to climb the mountain, establish the OP, monitor the action below, and call for fires if necessary. 21 June 2006 was the planned date for the GOWARDESH THRUST main effort movement into the valley. The *kill team* would begin its move on 18 June, infiltrating by climbing up to their planned observation position on the ridgeline which was located just three miles northwest of the village of Gowardesh.

On the eve of the operation, the enemy situation in the vicinity of the Gremen Valley was largely unknown to the task force. The 3-71 CAV staff utilized what limited information on the enemy it could gather based on interviews with US Special Forces detachments that had operated in the region previously and interviews with the local population. Being the first battalion-sized American element to establish a presence in the area, the squadron had no existing pattern of analysis for the enemy. According to Captain Ross Berkoff, the squadron's Intelligence Officer (S2), "the historical data necessary for charting trends in enemy response time, organizational capabilities, aggressiveness, and marksmanship" was simply not available. Therefore, it was uncertain what kind of opposition TF *Titan* elements, including the *kill team*, would face as they pushed into the Gremen Valley. Based on what little information was available,

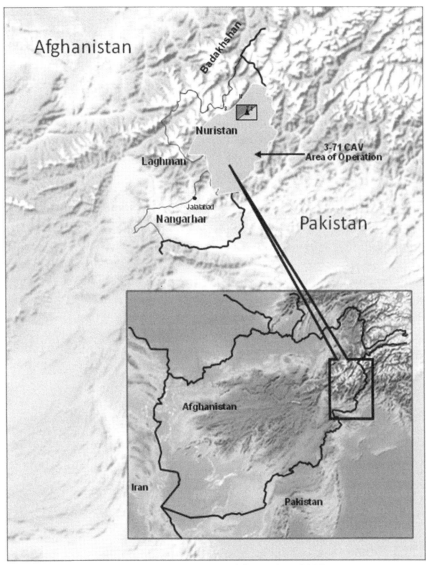

Figure 2. TF *Titan* Area of Operation, 2006.

Captain Berkoff determined that the enemy's most likely course of action would be to simply observe the Americans, to include the *kill team*. An organized enemy attack on the *kill team* was deemed the most dangerous course of action but was seen as much less likely.[6] The squadron staff planned for the *kill team* to have indirect fire support from 105mm howitzers and 120mm mortars during their climb and while in position and anticipated that this would be sufficient to counter any unexpected engagement with the enemy. As long as the staff's assessment of enemy intent was correct, the *kill team* had the right size, composition, and fire support to make the mission a success. As one of the members of the squadron's battle staff later claimed, "From the aspect of planning and not knowing, based on the INTEL that we had, that was the best plan we could come up with."[7] Although the staff did not know it at the time, this plan would soon face a significant test.

Burdened with double the normal combat load which included just enough food and water for the three-day infiltration up the mountain, the 16 Soldiers were trucked from FOB Naray near the village of Kamu to a task force mortar position near the Gowardesh Bridge, located just south of the Gremen Valley. Each team member carried more than 50 pounds of equipment with them, not including their personal weapon. Early on the morning of 18 May, the team began its walk up the ridgeline toward Mountain 2610. The Soldiers knew they had three days to get into position. They did not know however, exactly what awaited them up on the mountain. For Staff Sergeant Cunningham, the mission was clear but prospects for success were uncertain. "Something didn't feel right about it," he said later.[8] The infiltration plan did not worry him. Cunningham was instead slightly apprehensive about the plan to extract the team once the mission was complete. "I don't like going into a place where I don't have a way out," he said later, "and I don't like having to come out the way I came in." Two courses of action had been developed for the *kill team*'s extraction. The first entailed airlifting the team out once a landing zone (LZ) was created on the ridgeline by advancing TF *Titan* elements during GOWARDESH THRUST. The second course of action involved the team walking down the mountain using a different route, one that ran directly into the village of Gowardesh.[9] Both of these courses of action were contingent TF *Titan*'s movement into the valley going according to plan. All Cunningham and the other members of the *kill team* could do was hope that it did.

The Climb

The *kill team*'s ascent over the next three days was typical of the difficult ground maneuver in northeastern Afghanistan. Navigating the steep, rocky terrain while carrying a combat load and enough provisions for 72 hours was challenging enough but the rapid increase in altitude only added to the difficulty and prevented the team from moving quickly. The squadron's actions during Operation MOUNTAIN LION had introduced the Soldiers to the challenges offered by the mountainous terrain so the physical trials came as no surprise. This familiarity however, did not make the task any less demanding for the *kill team*. Indeed, in compiling their end of tour after action review (AAR), the Soldiers of the 3d Brigade emphasized the severe effects of the physical environment on their operations:

> The entire Kunar/Nuristan experience was impacted by terrain and weather. Heat injuries [occurred] due to the weight of full kit and foot movement over rugged terrain. We were also affected by cold weather or exposure injury due to operations in a mountainous environment. If it wasn't the heat it was the mountains, if wasn't the mountains it was the cold and or rain and snow.[10]

The AAR did not exaggerate. The terrain was so treacherous that the brigade suffered significant Soldier injuries and even deaths from falls in mountains. Seeing fellow Soldiers die as a result of the terrain was not likely something the unit anticipated when it prepared for its deployment to Afghanistan.

In an effort to negate the effects of the early summer heat, the *kill team* made frequent stops on their climb up the ridgeline of Mountain 2610 and moved during the night as much as possible. They took both short and long halts, taking time to observe the valley below and keep a watch for enemy in the area. During a number of these halts the *kill team* began to detect a significant amount of movement in the valley. "We observed a bunch of guys going up and down…huge groups of guys," Staff Sergeant Cunningham recalled, "Things just didn't seem normal, all of these people are moving into and out of the valley."[11] They reported these findings to the task force HQ over the radio and continued their climb. When they finally arrived at their planned position on the ridgeline it was after dark on the night of 20 June. Even with the lack of light, it was clear to them that they were not the first people to occupy that spot. Evidence of heavy foot traffic, perhaps recent, marked the site. To the east, what appeared to be a wide, well-used footpath ran along the ridge and dipped down toward the valley below. "This ain't good," Cunningham remembered thinking. The task of remaining undetected suddenly seemed much more challenging. The team radioed back their findings again and prepared to establish a perimeter.

Despite the evidence of foot traffic, the position they now held was suited to accomplishing the mission. They were perched atop a bump in the ridgeline with relatively flat ground immediately around them with a number of trees and rocks for cover and concealment. To the east they had a good view of the Gremen Valley to include a number of the operation NAIs. One of these was the suspected residence of a well-known insurgent leader, Hadji Usman, who was high on the Task Force target list. The position of the patrol base was such that the portion of the valley directly below was difficult to observe as it stretched further to the north. The *kill team* could identify suspected insurgent trail networks. For the observation requirements set forth in the task force mission, the *kill team* was where it needed to be. According to an official Army account of the engagement, "The position the patrol now occupied was in a known enemy sanctuary and near the Urstan Pass, a major trafficking route used to smuggle foreign fighters, weapons, IED [Improvised Explosive Device] making materials, and other supplies into the region from Pakistan."[12] If the task force sweep during GOWARDESH THRUST caused any movement in the valley, the team would likely see that movement from where its Soldiers were now standing.

Directly above the team's position, the mountain ridge continued its upward slope to the north and a group of large rocks with a small clearing beyond made up the southern boundary of the position. On either side of the ridge, the ground dropped off at a sharp angle. As such, the *kill team* was faced with the typical tactical constraint of a ridgelines such as that if they needed, they could only maneuver in two directions, up the ridge to the north or back down to the south. While not an ideal situation, the terrain offered no viable alternative. As most US Soldiers conducting operations in Kunar and Nuristan would learn, the terrain was seldom if ever ideal. Perhaps thankfully, the *kill team*'s time in this position was supposed to be short. The original plan for the 16 Americans now taking up positions in the dark was for them to begin their overwatch within the next hours. As a result, they did not immediately begin to prepare deliberate fighting positions or conduct other preparations associated with a lengthy static defensive position. Another factor contributing to their actions that evening was physical exhaustion. The three day climb in temperatures often in excess of 100 degrees, had taken its toll and most of the Soldiers were almost out of food and water. For the remainder of the night, the team recovered from the movement up the ridgeline and kept watch on the NAIs in the valley below.[13]

Before daybreak on 21 June, the *kill team* received the news that Operation GOWARDESH THRUST was on a three day hold. TF *Titan* would not be entering the Gremen Valley that day. The night before, a convoy from Charlie Company had suffered an IED attack while traveling to a meeting with local officials at a border crossing. A number of Soldiers were wounded and the vehicle sustained significant damage. The attack immediately raised questions about the security of the main route of advance into the Gremen Valley. The 3-71 CAV leadership decided to delay the operation in order to conduct route clearance and to inspect the counter-IED equipment on all the squadron's vehicles.[14]

When Staff Sergeant Cunningham learned of the delay, he remembered thinking to himself, "That is not good."[15] Indeed, it left the team in an uncertain position. Most of the Soldiers were in significant need of food and water and what little they had left would not last another 24 hours. Originally, the squadron planned to deliver an aerial resupply package to the *kill team* in conjunction with the air assault into the valley scheduled for 21 June with the assumption that the amount of air activity over the area would mask a single UH-60 dropping off supplies to the team. Now, a separate aerial resupply would have to be conducted during the day which would more likely alert the enemy to the team's location.

Perhaps more importantly, the 16 Americans would need to find a way to remain undetected by the enemy for three days. During Operation ANACONDA in March 2002, US Special Operations forces were deployed in a similar fashion to provide overwatch of the Shahikot Valley in Afghanistan. Those teams however, were smaller, highly-trained, and well-equipped for such a mission. Even with their training, Special Operations observation teams were sometimes discovered by insurgent enemies, as was the case with a four-man SEAL team during Operation RED WINGS in Kunar Province almost a year prior. Once discovered by a group of local Afghan goat herders, the SEAL team decided to release them rather than hold them captive or kill them. Less than two hours later, the team was attacked by a large force of Taliban fighters and all but one of the SEALs were killed. There was no question in the mind of the surviving team member that the Taliban fighters had been informed of their presence by the goat herders.[16] The Soldiers within the TF *Titan kill team* on Mountain 2610 were not Special Operations Soldiers nor were they well equipped for an extended stay in the mountains. These facts were not lost on the members of the team and according to at least one report, several of the Soldiers "had gut feelings that something wasn't right."[17] Nevertheless, the word came from FOB Naray that the team was to remain in position. The task force staff reached this decision after considering the fact that the *kill team* had immediate fire support available from the 105mm howitzers and 120mm mortars in case of an enemy engagement. Another critical factor was the larger operational schedule. GOWARDESH THRUST was imminent and there was not enough time to pull the team off the mountain only for it to turn around immediately and hike up the mountain again. Given these considerations, the staff was "pretty comfortable" with the decision to leave the *kill team* in place.[18] It was the first of three important decisions that would impact the *kill team* that day.

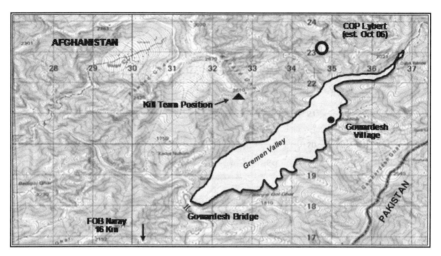

Figure 3. Gremen Valley Area including Mountain 2610 and Kill Team Position.

By 0800 on 21 June, it was clear to the members of the *kill team* and to the TF *Titan* staff that the team needed resupply of food and water that day. Although the resupply posed an increased risk of enemy detection, the team's critical need for water was deemed paramount and the task force's leadership felt it had little choice but to conduct the resupply as soon as possible.[19] This was the second important decision that day. The planned delivery point for the aerial resupply was several hundred meters to the northwest of the patrol base. When the *kill team* received word that a UH-60 was inbound with resupply, the team leaders split up the group of 16, taking the majority of them to the resupply landing zone (LZ) to provide security and carry the water and food back to the patrol base. The remainder of the team stayed at the patrol base and watched the valley.

At approximately 1330 hours, a single UH-60 flew in low over the Gremen Valley, directly over one of the suspected enemy structures that the *kill team* was monitoring. Cunningham could barely believe it. "This [pilot], he didn't talk to me, he just came around…right over the bad guy's house," he recalled, "He flew right to me, stopped right above me, and dropped the stuff."[20] The squadron staff had initially planned for the resupply UH-60 to approach Mountain 2610 from a different azimuth and to fly over the delivery point and drop the resupply bags without stopping. This would make it more difficult for any enemy watching to pinpoint the *kill team*'s whereabouts. "That didn't happen," the squadron fire support NCOIC later noted, "The aircraft hovered over them, which we think gave away their positions. We are still upset about that."[21] Once the helicopter deposited the large white bags of food and water, it departed for FOB Naray. The weather on the ridge was clear, permitting full view of the UH-60 from the ridgeline above and valley below. The drop had actually occurred very close to the *kill team*'s patrol base rather than at the designated LZ. This fact made a number of the Soldiers uneasy.[22] The Soldiers at the drop point gathered the food and water bags and began moving them the short distance back to the base.

Meanwhile, Specialist Max Noble, who was pulling security at the team observation point, identified what appeared to be an Afghan male watching the resupply operation through a pair of binoculars. The man was located approximately 1,500 meters to the east in the valley and was intently observing the Americans' location. Minutes later, the majority of the team arrived back at the patrol base with the resupply. Specialist Noble, the team medic, alerted Staff Sergeant Cunningham to the observer in the valley. Both Cunningham and Monti watched the man for several minutes as he continued to observe their position on the ridgeline and then disappeared. "He was clearly looking at us," Cunningham recalled.[23] Although there was the unlikely chance that the local man was not a threat, one thing that seemed clear to the Soldiers on the ridge was that the *kill team*'s position had been compromised.

The Fight

Monti and Cunningham now faced a tough tactical decision of whether to move or stay in position. The dilemma was strikingly similar to that of SEAL team during Operation RED WINGS. They believed their position had been compromised, but it was uncertain what that would mean for their mission. For the SEAL team, discovery by a local Afghan had led to an attack by a large enemy force. Conventional tactical wisdom might have suggested finding a new location for the patrol base but there was nothing conventional about conducting operations on a narrow ridgeline in the mountains of Afghanistan. Back at FOB Naray, the TF *Titan* battle staff was not overly concerned with the situation. After all, the intelligence assessments had predicted that the enemy would indeed attempt to watch the *kill team* while suggesting that a direct attack was less likely. Thus, the team's situation had been anticipated to a degree. The TF S2 recalled "no discussion" within the staff on moving the team to a new location at that point.[24] During the intense planning and coordination for the upcoming operation, the team's report of the Afghan with the binoculars was not interpreted as an emergency at FOB Naray.

Up on the mountain, the *kill team* leaders discussed their tactical options. Staff Sergeant Cunningham later described the conversation between himself and another team NCO, Sergeant John Hawes, regarding their situation:

> We can't go any further up [the ridgeline]. If we go any further we're possibly going to get ambushed on the way up there. So we're either going to hold this position or move back down. If we move back down, we're going to give them the high ground...so we decide to hold the high ground, pull security on what we're supposed to be watching.[25]

In addition to the possibility of being attacked, there was no assurance that the team would find a vantage point further north up the ridge or further down it that would allow them to accomplish their mission.

Given this uncertainty, Staff Sergeant Cunningham and Staff Sergeant Monti decided that the team would stay in position. Unaware of the large group of insurgents that was moving toward them from the north as the late afternoon turned into evening, Monti and Cunningham informed the rest of the team that they would remain where they were and directed them to continue their observation of the NAIs in the valley. The Soldiers cross-leveled supplies, began guard shifts, and took time to eat and drink. The decision to remain at the patrol base was the third of three fateful decisions leading up to the deadly engagement.

Six of the *kill team* members were positioned at the northern side of the patrol base, facing a line of thick vegetation. Three Soldiers continued to monitor the valley on the eastern side of the position. To the south, the team leaders formed what they called an "Alamo," or "a place you can fight back to," near a group of large rocks and trees.[26] There, Cunningham, Monti, and Hawes continued to discuss the security plan for the night and the next day. Although the members of the *kill team* were alert and vigilant, they had no indicators of enemy activity in the vicinity of their position.

At 1845, almost five hours after the resupply UH-60 departed the ridgeline, Specialist Franklin Woods suddenly heard the sounds of people approaching through the wooded area to the north of the patrol base and seconds later, a heavy barrage of Rocket Propelled Grenade (RPG) rounds and machine gun and small arms fire burst from the wooded area. Round after round slammed into the ground and rocks around the patrol base, sending the Soldiers scrambling for cover. It was clear from the sheer amount of fire striking the position that the attackers were numerous and very well equipped. Staff Sergeant Cunningham later recalled the intensity of the initial barrage by stating, "I've been in a bunch of firefights. That was probably one of the most intense."[27] Specialist Daniel Linnihan remembered being amazed by the "ridiculous" volume of RPG fire. Now well aware that the enemy group maneuvering against them was not just a few insurgents with AK-47s, but something much larger, Linnihan recalled, "They clearly came up there to kill every one of us."[28] After several minutes of intense fire coming from the woods to the north, it became clear to the Soldiers on that end of the perimeter that approximately 50 insurgents were coming toward them and attempting to flank the patrol base. The enemy had established two support-by-fire positions to the north and northwest and continued to pour machine gun fire into the patrol base.

After the initial shock of the sudden attack wore off, the Soldiers of the *kill team* began to fight back. Those positioned closest to the enemy advance on the northern edge of the perimeter returned fire into the tree line but realizing that they were too exposed, began to fall back to the "Alamo" and the cover of the rocks. Specialist Shawn Heistand fired a burst from his assault rifle in the direction of the enemy and then got up and darted for the southeast corner of the team's position. Private First Class Brian Bradbury, a squad automatic weapon (SAW) gunner, followed right on Heistand's heels but was struck by enemy fire almost immediately and went down in the clearing badly wounded. The three other team members on the northern end of the perimeter sought to put some distance between themselves and the enemy. In unison, they rushed to the southeast, diving behind the cover of the rocks as small arms fire and RPG rounds whizzed by. In the dash, Private First Class Mark James was wounded but was able to make it to the rocks, where Specialist Matthew Chambers grabbed him and began to treat his wounds.[29] The fight was just two minutes old and the Americans already had two Soldiers wounded, one of them critically and out in the open.

While Soldiers were repositioning themselves at "Alamo" at the southern end of the perimeter, Sergeant Patrick Lybert laid down a continuous cover fire with his assault rifle. The enemy fighters could be heard shouting to one another in the woods ahead as they tried to maneuver closer to the Americans. Lybert continued to fire steady bursts at the sounds of the enemy, trying to buy time for the other members of the *kill team* as they formed a new perimeter behind the rocks. Minutes later, Sergeant Lybert was struck by enemy fire which mortally wounded him. Other members of the team saw Lybert's head slump forward, and when they called to him, he did not answer or move. The *kill team* had suffered its first fatality.[30] Initial attempts to get to Lybert were repulsed by the steady stream of incoming fire. Specialist Linnihan, who was positioned closest to Lybert, could not get to him but during one of his several attempts, he was able to secure Lybert's weapon and hand it to one of the other Soldiers behind the rock barrier.

As the *kill team* quickly reformed a tight perimeter to the south of the rocks, Staff Sergeant Cunningham directed the team's fire against the enemy while Staff Sergeant Monti initiated a call for fire over the task force fire support radio network. In his situation report, he expressed the severity of the team's situation and the possibility that it might be overrun. The fire support coordinator on the other end of the radio remembered Monti's "very calm, very clear" voice on the radio despite what was an increasingly grave situation.[31] In his normal measured tone, Monti provided the grid coordinates for mortars and artillery support, specifying the need for "danger close" fires or rounds that would land approximately 200-300 meters from the team. One thing that many Soldiers operating in northeastern Afghanistan would learn is that the steep inclines in the mountainous terrain greatly influenced the effectiveness of indirect fire. Where one might be afforded a little leeway in coordinating fires for a relatively flat target area, there was no such leeway in the mountains. Calculation errors of only a few meters could render an artillery or air strike completely ineffective. Within seconds, task force mortar fire began slamming the woods to the north but further off than the team needed. Unknown to the Soldiers of the *kill team*, the crews manning the 120mm mortar position at Gowardesh Bridge came under attack at almost the same time as the Soldiers on the ridgeline, albeit from a smaller enemy force. While fighting off the attack on their own position, the mortar platoon at the bridge continued to answer Monti's call for fire. As the mortar rounds landed in the trees, enemy RPG and RPK fire decreased. That fire however, was still intense enough to keep the Soldiers pinned to the ground and unable to maneuver or get to Sergeant Lybert or Private First Class Bradbury.

At the TF *Titan* tactical operations center (TOC) at FOB Naray, the squadron staff was reacting to the initial reports of contact with the enemy. Two units were simultaneously defending themselves from unexpected insurgent attacks. This was the first indication that the enemy operating in Nuristan was much more sophisticated, organized, and aggressive than previously thought. For Staff Sergeant Cunningham, the coordinated attacks proved that the enemy consisted of "pretty smart guys."[32] As the radio reports of the engagement continued, not much could be gathered about the composition or the exact disposition of the enemy. Staff Sergeant Monti was pinned down by the enemy's RPGs, unable to provide the details the squadron staff wanted. It was clear to those in tactical operations center or TOC however, that the *kill team* was up against a sizable enemy force. The squadron S2 recalled, "We were surprised with the number of enemy fighters reportedly involved since it constituted the most dangerous course of action."[33]

The first reaction of many in the TOC was to assume that the helicopter resupply had indeed alerted the enemy to the *kill team*'s presence allowing them to organize an attack. Given that the attack occurred approximately five hours after the resupply drop, it is also possible that the enemy force was already positioned on the mountain and aware of the Americans' location when the UH-60 hovered over them. The coordinated attack against the mortars also supported the notion that what the *kill team* was facing was much more of a planned organized attack than a rapid reaction to the helicopter sighting. As the squadron S3, Major Timmons, commented, "You don't consolidate that number of people with that much armament that fast without significant preparation."[34] In all likelihood, the enemy force of 50-70 fighters was already moving toward the *kill team* planning an attack. The resupply mission of that afternoon might have served the insurgents as confirmation of the kill team's location. If true, the enemy had likely observed the kill team as they entered the valley near the Gowardesh Bridge and began their infiltration. Thinking back on the event, Captain Berkoff made an apt observation, "Since we underestimated the enemy's organizational capabilities, we should certainly not underestimate their degree of situational awareness on their own mountain tops."[35] Whether the attack on the *kill team* was triggered by observation of the team near the bridge or by the aerial resupply remains unclear. What is evident however is that the enemy in Nuristan in 2006 was capable of maneuvering a large well-equipped force to within a hundred meters of the team's position undetected and launching a coordinated attack that took both the team and the task force by complete surprise.

While continuing its supporting RPG and machine gun fire and perhaps spurred on by the incoming mortars, the enemy began to attempt to flank the kill squad's position from both the northeast and western sides. To the northeast, Private Sean Smith had been able to secure a sniper rifle while falling back to a position behind the rocks and now used it with great effect. In a spot hidden from enemy view, he engaged a number of insurgents trying to use the footpath to skirt the eastern side of the team's position, killing a number of them. Remembering Smith's effective

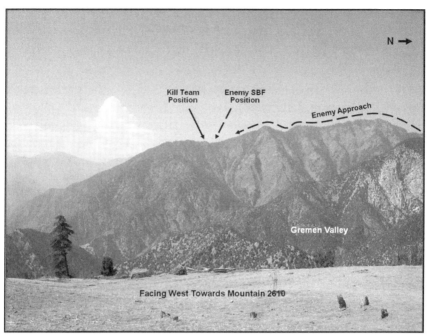

Figure 4. View of Mountain 2610 with US *Kill Team* and Enemy Support By Fire Position Shown.

defense of their eastern flank, Staff Sergeant Cunningham recalled, "Every guy that would come across that trail would get dropped."[36] Meanwhile, Monti maintained communications with the mortar platoon in an attempt to adjust the fires closer to his position. He paused long enough to engage an advancing group of insurgents just 20 meters away with his M4 carbine and a grenade, forcing them back. The *kill team* fought in three directions, directing fire to the north, east, and west. For the moment, the enemy's flanking attempts had been checked but the steady insurgent fire from the woods continued. Adjusted rounds from the mortars began to hit their mark to the north but the enemy persisted. Outnumbered and outgunned, the Americans needed more firepower to turn the tide.

At this time, Staff Sergeant Monti and Sergeant Hawes turned their attention to the wounded Private First Class Bradbury who lay approximately 60 feet away in the open. Although partially hidden from enemy observation, Bradbury was still very much in danger of being hit with indirect fire. He could weakly communicate with the others behind the rocks but his wounds were such that he could do little else. Staff Sergeant Monti quickly decided that he had to retrieve Bradbury and get him to

Figure 5. Scheme of Maneuver during Firefight Above Gowardesh.

safety behind the rocks.[37] He turned control of the squad radio over to Sergeant Christopher Grzecki and prepared to make a rush into the open. His first attempt carried him a few feet away from the rocks before a hail of enemy small arms fire forced him to take cover behind a small rock wall skirting the edge of the perimeter. Here, Monti found himself next to the body of Sergeant Lybert. After waiting another minute, he jumped over the wall and tried again to make his way to Bradbury only to be forced back again by withering fire. Before making what would be his final attempt to save Bradbury, Staff Sergeant Monti yelled to the other members of the *kill team* that he needed covering fire from their M203 Grenade Launcher. Sergeant Hawes fired a number of 203 rounds into the tree line to the north and the rest of the team opened up with a barrage of supporting rifle fire. At this point, Monti sprinted into the open toward Bradbury and within seconds was mortally wounded when an RPG round slammed into the ground a few feet from him. Minutes later, just as the sun dropped behind the ridgeline, Staff Sergeant Monti died from his wounds.

At this point in the battle, the *kill team* was holding out for darkness and close air support both which arrived simultaneously. A B-1 bomber circling overhead delivered a number of 500 and 2,000 pound Joint Direct Attack Munitions (JDAM) which slammed into the woods on the northern stretch of the ridgeline. The bombs shook the ground and covered the team's position in a cloud of dust. Numerous 105mm artillery and mortar rounds also found their mark in the woods, forcing the enemy fire to wither. One squadron Fire Support Officer (FSO) remembered, "We used nearly every mortar round they had out there."[38] The *kill team* emptied their magazines in the direction of the attack, although in the increasing darkness, they were relying more on sound than sight to target the enemy. Less than ten minutes later, the enemy fire ceased and only the sounds of circling aircraft and American voices shouting to one another could be heard. The firefight was over. Two noncommissioned officers from the team were dead and another Soldier was critically wounded but alive. Later estimates would place the number of enemy killed at 26 and the number of wounded at 17. The entire engagement lasted less than 60 minutes.[39]

Aftermath

The fight on Mountain 2610 may have been over but the members of the *kill team* would not know it until the next morning. They spent the rest of night of 21-22 June on 100 percent security anticipating another attack. Nervousness and fatigue combined with the painful grief at the loss of their two fellow Soldiers to make the wait for sunrise a harrowing task. While many of the Soldiers would have taken the chance to withdraw from the mountain in the dark, to do so with two KIAs and one critically wounded Soldier presented too much of an obstacle. Immediately after the fight, a medical evacuation aircraft (a UH-60 Black Hawk) from the 159th Medical Company flew up to the ridgeline to extract the wounded. Tragedy struck the team once more as Bradbury and a medic, Staff Sergeant Heath Craig, were being hoisted up to the helicopter. The litter basket in which they sat began to oscillate rapidly causing the winch to pinch the lifting cable against the side of the helicopter. As a result, the cable snapped, sending Bradbury and Craig plummeting hundreds of feet to the ground below. Both were killed.[40] It was another devastating blow to the team as it sought to regroup from the jarring events of that evening.

As much as the battle staff of TF *Titan* wanted to get the *kill team* off the mountain that night, the proposition seemed too risky, particularly after the MEDEVAC accident. It is also important to note that just six weeks prior, on the night of 5 May 2006, the task force suffered a catastrophic crash of a CH-47 that was extracting Soldiers from a small landing zone perched on a ridge above the Chawkay Valley. Ten Soldiers were killed, including the 3-71 CAV Commander Lieutenant Colonel Joseph Fenty. The emotional and psychological toll of the crash still weighed heavy on the squadron. Major Richard Timmons, the Squadron S3 stated, "People

18

were very hesitant to place helicopters in dangerous positions" after the accident.[41] Captain Berkoff agreed, remembering that "No one wanted to risk another bird."[42] The 3-71 Commander Lieutenant Colonel Michael Howard, reported the decision to leave the *kill team* in position until the next day up to the 3rd BCT and CJTF-76. All agreed that attempting a helicopter extraction that night was simply too dangerous.[43] Instead, the task force ensured that they had continuous aerial surveillance and fire support coverage for the team's position for the duration of the night.

When morning broke on 22 June, the *kill team* was more than ready to return to FOB Naray. They had fought hard against a determined enemy that outnumbered them and beat them back, but nearly 25 percent of the team did not make it off the ridgeline alive. It was not a definitive defeat, but also not a victory of any kind. The team made a quick descent from the mountain during the day on 22 June and was back at FOB Naray before nightfall. Fellow TF *Titan* Soldiers met the tired members of the team as they entered the main gates of the FOB and assisted them in removing and securing their equipment. The story of the fight spread through TF *Titan* and Staff Sergeant Monti's heroism quickly became apparent to all in the command. That day, the task force commander initiated Monti's posthumous promotion to the rank of Sergeant First Class.[44] Two days later, a patrol from the squadron's Charlie Company returned to the site of the firefight to secure the small amount of tactical gear that the *kill team* was forced to leave behind. One of the officers on the patrol expressed his amazement at the amount of blood present at the site, uncertain if it was enemy or friendly. He later commented to a friend, "If death had a smell that was it."[45] No enemy remains were discovered at the site. The squadron pushed into the Gremen Valley on 24 June, establishing a presence that they would not relinquish. From September to November of 2006, TF *Titan* manned a combat outpost (COP) on a ridgeline just east of Mountain 2610 and named it in honor of Sergeant Patrick Lybert. From there, the Soldiers of TF *Titan* built relationships with the local population and monitored activity in the valley for the remainder of their deployment.

Over three years later, Sergeant First Class Monti was awarded the Medal of Honor, the second recipient of the award from the war in Afghanistan. Indeed, the story of Monti's performance under fire dominated the news of the Gowardesh fight in the years following. The events of that night however, offered a number of insights for US forces fighting in Afghanistan well beyond the acts of a single individual. When Sergeant Daniel Linnihan was asked what advice he would offer to other Soldiers based on his experience on Mountain 2610 that night, he stated simply, "complacency kills."[46] The attack had occurred so suddenly, that according to Linnihan, it served as a sobering reminder that insurgent ambushes, particularly in the mountains of northeastern Afghanistan, are rarely detectible. "I never took my gear off again," for the remainder of the unit deployment, he later claimed.[47] For Staff Sergeant Cunningham, the enduring lesson was simple. "More vigilance" was needed in dealing with an enemy that was both elusive and bold.[48] This message would prove prophetic two years later in a spot that was immediately adjacent to the scene of the Gowardesh firefight. The units that replaced TF *Titan* in the Gremen Valley area continued to man COP Lybert and experienced little to no enemy activity for months at a time. In May of 2008, a *Stars and Stripes* reporter visiting US Soldiers operating near Gowardesh reported, "The chances of a well-coordinated and successful [enemy] attack are pretty slim... The biggest enemy they [US Soldiers] face, at least for the moment, is the boredom."[49] Two months later, Soldiers from the 3d Brigade Combat Team, 1st Infantry Division were enjoying what they perceived as a low-threat environment at COP Lybert, with one Soldier describing the mission as "a normal day-to-day life" and others expecting "a quiet deployment." On 11 September 2008, insurgents attacked the COP without warning and a two-hour firefight ensued. In the end, support from Apache helicopters and A-10 aircraft was needed to drive off the attackers, leaving one US Soldier dead.[50]

If nothing else, the firefight of 21 June 2006 demonstrated that US forces could easily underestimate the enemy. The insurgents were more capable and thus more dangerous than the US Soldiers had anticipated. After his deployment, Captain Berkoff, TF *Titan*'s Intelligence Officer, made this point emphatically:

> As a whole, we clearly underestimated the enemy's organizational capabilities. This particular engagement taught us that the enemy in Nuristan did not mind fighting us during the night. They were opportunistic fighters. They launched that attack with little prior planning and little thought of their own risks. It was the first time they ever saw American Soldiers set up in their own woods/backyards and they were eager to do some damage.

With this type of aggressive, organized, and motivated adversary operating around them, the lessons of "complacency kills" and "more vigilance" offered by the survivors of the engagement seem fitting indeed. It is unclear what role, if any, tactical complacency played in the events surrounding the engagement on Mountain 2610 but what is evident is that the Soldiers of the *kill team* and TF *Titan* emerged from the event much more aware and prepared for the type of threat they faced.

The firefight of 21 June 2006 also highlights the critical role played by artillery and close air support in operations in northeast Afghanistan. The Soldiers of the TF *Titan kill team* were the first Americans to fight in the Gremen Valley but the outcome of the engagement was similar to other engagements in Nuristan and Kunar provinces. In the Korengal, Waygal, and Chawkay valleys, direct engagements with the enemy usually required both artillery and close air support to disrupt an insurgent attack. The extreme terrain usually prevented the basic tactical principles of identifying, closing with or even pursuing the enemy. Often, the exact location of the enemy was impossible to ascertain from the ground. As such, the Soldiers of TF *Titan* and those that arrived after 2006 came to rely heavily on friendly aviation assets and aerial surveillance platforms. If they did not arrive already prepared to coordinate fire support missions, the Soldiers operating in northeastern Afghanistan would have to become adept at directing artillery and close air support. This skill was critical even for Soldiers at the squad level.

The logistical constraints of operating in northeast Afghanistan were also made painfully apparent as the *kill team* found it stuck on the ridgeline in need of food and water, forcing the task force to conduct a high-risk resupply operation in daylight. The act of climbing mountains, some in excess of 7,000 feet, brought with it a myriad of limitations in relation to the time, flexibility, and contingency of any operation. In the case of Operation GOWARDESH THRUST, when the mission was delayed, the squadron staff believed the only feasible option was to leave the *kill team* in position primarily because of the time and effort required to infiltrate to the position. After being accustomed to a modern training world where HMMWVs, trucks, and helicopters facilitated rapid movement on the battlefield, the Soldiers of TF *Titan* found the terrain of Kunar and Nuristan to be a stark, tactically debilitating obstacle that slowed logistical operations from squad to division level to a plodding and frustrating pace. This realization came from the hard experience of climbing, step by step. An interview conducted with TF *Titan* Soldiers and officers towards the end of their rotation suggested that, "no one, it seems, told them they would have to fight a Vietnam-style war at high altitudes."[51] In such an environment, the type of food and water shortage experienced by the *kill team* on 21 June, something rare in the training world, became a more common occurrence throughout TF *Titan's* rotation that year. One squadron officer stated, "We've gone on mission after mission after mission where we've gone black on food and water," adding, "We end up stuck on a mountain top for two weeks. If you can't get us off a mountain, don't put us there."[52] These logistic and operational frustrations defined much of TF *Titan*'s year on the border between Afghanistan and Pakistan and would remain defining factors for follow-on units as well.

Despite the numerous challenges offered by the environment, the Soldiers and leaders of TF *Titan* found a way to wage a counterinsurgency in Kunar and Nuristan. The enduring message they offered their successors was that this war was very much a process of trial and error both tactically and logistically. It is true that the Soldiers of TF *Titan* were not totally prepared for the challenges that awaited them in the mountains. This fact is neither surprising nor unexpected given that it was the first American battalion to operate in the region. The firefight above Gowardesh was one of the first of many poignant examples of American Soldiers facing significant adversity in the mountains of northeast Afghanistan and fighting through it. When they were attacked by an enemy with superior numbers and sufficient firepower, the members of the *kill team* stood their ground on challenging terrain and fought off the attack. They did not withdraw or permit themselves to be overrun. They were able to successfully integrate artillery and close air support into the fight, a critical task in this war. Although a number of tactical decisions could have been made differently, it is impossible to know how different actions would have changed the outcome of the engagement, if at all. The events of that night above Gowardesh have up to the publication of this study served as the context for one of the war's few Medals of Honor. The greater significance of the firefight of 21 June 2006 however, lay in its indication of the campaign to come. The units that followed TF *Titan* into northeast Afghanistan over the four years that followed their deployment would continue to face serious difficulties created by the unforgiving terrain, harsh weather, and a committed insurgent enemy. Ultimately, the story of TF *Titan kill team* on Mountain 2610 is the story of all US Soldiers finding a way to wage a complex counterinsurgency in a very challenging operational environment.

Notes

1. Michael A. Coss, "Operation Mountain Lion: CJTF-76 in Afghanistan, Spring 2006," *Military Review* (Jan-Feb 2008): 23.

2. Company Operations History for June 2006, Charlie Company, 3d Squadron, 71st Cavalry Regiment, dated 15 July 2006; Company Operations History for June 2006, Headquarters and Headquarters Troop, 3d Squadron, 71st Cavalry Regiment, dated 15 July 2006.

3. Associated Press, "U.S. Forces Push further into Afghanistan," 8 August 2006. www.military.com/NewsContent/0,13319,109103,00.html. (Accessed on 13 March 2011.)

4. Lieutenant Colonel Richard Timmons, Interview by Lieutenant Colonel Clay Mountcastle, 12 April 2011. Timmons served as the Squadron XO and S3 during the unit's deployment to Afghanistan in 2006-2007.

5. Official US Army Narrative, Congressional Medal of Honor for SFC Jared C. Monti. Accessed at www.army.mil/medalofhonor/monti/narrative.html on 21 February 2011; Zimmerman, Dwight J. and John Gresham, *Uncommon Valor: The Medal of Honor and the Six Warrior who Earned it in Iraq and Afghanistan* (New York: St. Martin's Press), 250.

6. Captain Ross Berkoff, Interview by Lieutenant Colonel Clay Mountcastle, 15 April 2011.

7. Sergeant First Class Hunsacker, Interview by Douglas Cubbison, Operational Leadership Experience (OLE) Project Team, Combat Studies Institute, Fort Leavenworth, KS, 24 January 2007, 7.

8. Staff Sergeant Christopher Cunningham, Interview by Douglas Cubbison, Operational Leadership Experience (OLE) Project Team, Combat Studies Institute, Fort Leavenworth, KS., 13.

9. Cunningham, interview, 13-14.

10. 3d Infantry Brigade Combat Team, 10th Mountain Division, US Army, TF Spartan AAR, May 2007, 9-30 to 9-31.

11. Cunningham, interview, 14.

12. Official US Army Narrative, Medal of Honor for SFC Jared C. Monti.

13. Monti MOH narrative; Interview with SGT Daniel B. Linnihan, conducted by author on 22 March 2011.

14. Timmons, interview.

15. Timmons, interview; Cunningham interview, 14.

16. Marcus Luttrell and Patrick Robinson, *Lone Survivor: The Eyewitness Account of Operation Red Wing and the Lost Heroes of SEAL Team 10* (New York: Little, Brown, and Company), 200-220; Andrew North and Tony Allen-Mills, "Downed US SEALs may have got too close to Bin Laden," *The Sunday Times* (London), 10 July 2005.

17. Elizabeth M. Collins, " 'Soldiers' NCO' earns Medal of Honor for heroic deeds in Afghanistan," Army News Service, 14 September, 2009. www.us.army.mil/news/2009/09/14. (Accessed on 20 March 2011.)

18. Timmons, interview.

19. Zimmerman, *Uncommon Valor*, 252; Army CMH narrative for SFC Monti.

20. Cunningham interview, OLE, 14.

21. Hunsacker interview, OLE, 13.

22. Collins, " 'Soldiers' NCO' earns Medal of Honor," 2. The author reported that SGT Hawes "was chagrined to see large white packages dropping less than 150 meters from them in broad daylight."

23. Monti MOH Narrative; Cunningham interview, 14.

24. Berkoff, interview.

25. Cunningham, interview, 15.

26. Cunningham, interview, 14.

27. Cunningham, interview.

28. Sergeant Linnihan, Interview by Lieutenant Colonel Clay Mountcastle.

29. Zimmerman, *Uncommon Valor*, 255.

30. Monti MOH narrative; Zimmerman, *Uncommon Valor*, 258.

31. Hunsacker, interview, 13.

32. Cunningham, interview, 16.

33. Hunsacker, interview, 14; Berkoff Interview.

34. Timmons, interview.

35. Berkoff, interview.

36. Monti MOH narrative; Interview with Cunningham, 17.

37. Zimmerman, *Uncommon Valor*, 260.

38. Interview with First Lieutenant Jorgensen, Interview by Douglas Cubbison, Operational Leadership Experience (OLE) Project Team, Combat Studies Institute, Fort Leavenworth, KS., 22 January 2007, 13.

39. TF Spartan AAR, May 2007, 6-17. The AAR inexplicably described the enemy attack as "lasting 18 hours," but all other accounts suggest it lasted between 25-40 minutes.

40. Company Operations History for June 2006, Charlie Company, 3-71 CAV, dated 15 July 2006.

41. Timmons, interview

42. Berkoff, interview.

43. Timmons, interview; Berkoff interview.

44. Company Operations History for June 2006, Headquarters and Headquarters Troop, 3-71 CAV, dated 15 July 2006; Monti MOH Narrative.

45. Jorgensen, interview, 14-15.

46. Linnihan, interview.

47. Linnihan, interview.

48. Cunningham Interview, 18.

49. Drew Brown, "Soldiers Battle Boredom in Afghan Valley," *Stars and Stripes*, Mideast Edition, 4 May, 2008.

50. Michael Gisick, "Fateful Day brings Post Back to War's Reality," *Stars and Stripes*, Mideast Edition, 25 September 2008.

51. Matthew Cole, "Watching Afghanistan Fall." www.salon.com/news/feature/2007/02/27/afghanistan. (Accessed on 10 March 2011.)

52. Cole, "Watching Afghanistan Fall".

Ambushing the Taliban
A US Platoon in the Korengal Valley
by
Scott J. Gaitley

It was near twilight and they had just finished eating and saying their evening prayers. Gathering their weapons and personal items, the men departed the village of Chichal on what they believed to be an uneventful journey. As they strolled down the darkened mountainous trails, quietly chatting with one another, those leading the column suddenly heard a distinctive metallic click. Not knowing what this noise was, the column stopped abruptly. Little did they know, many of the men who began this nighttime journey would not make it home that fateful night.

On 10 April 2009, 2d Platoon, Bravo (*Viper*) Company, 1st Battalion, 26th Infantry (1-26 IN) "Blue Spaders," killed 15 insurgents in a modified linear-style ambush along the Sawtalo Sar Ridgeline in the Korengal Valley. At 1925 hours, 26 enemy personnel traveling down the Chichal/ Donga Trail, roughly two and a half kilometers east of the Korengal Outpost (KOP), unknowingly walked into the waiting firepower of 2d Platoon. This was an extraordinary and rare opportunity for the Soldiers to see their intended targets close-up and in such notable numbers. Members of the platoon were concealed and tactically positioned alongside the trail, providing them with superior fields of fire. The successful ambush enabled 2d Platoon to interdict a sizeable and heavily armed insurgent force moving freely throughout the Korengal Valley (see Figure 1).

Background

The Korengal Valley is a mountainous region approximately 10 miles in length and is comprised of narrow and towering rugged rock walls ascending to 3,000-7,500 feet in elevation which sharply descends toward the stream and its flood plain.[1] The Korengal Valley is considered one of the deadliest and most highly contested locations within the 14 provincial areas under the authority of Regional Command East (RC-East), the Coalition's military command in charge of operations in eastern Afghanistan. Located 25 miles west of Asadabad, the capital city of Kunar Province, the Korengal has earned the unpleasant reputation as the "Valley of Death" due to the 35 Americans killed there from 2005 to 2010.[2]

The Korengal stream courses through this arduous terrain in a northerly direction into the Pech River which runs east to west across the Pech

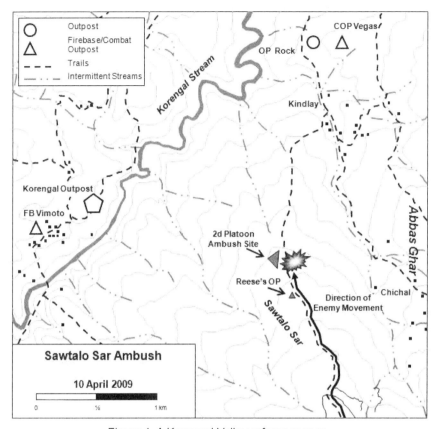

Figure 1. A Korengal Valley reference map.

District (one of 12 districts within Kunar Province). The population resides in villages randomly scattered near the valley floors and built among the steep mountainsides along the narrow valley. Areas between the villages are predominantly desolate and uninhabitable. For the villagers visiting and trading with one another, beaten dirt paths are the customary routes of travel. The region is inhabited primarily by Pashtuns from the Safi tribe who are mixed with ethnic Nuristanis. The dominant language spoken throughout the valley is *Korengali*, otherwise referred to as *Pashai*. Lacking a developed road system connecting the valley to other provincial areas, Korengal residents are isolated and normally distrustful of outsiders including other Afghans. The inhabitants of the Korengal are considered to be a fiercely inward society with a longstanding grievance against those trying to impose control over them. This attitude helps explain why Afghan forces and government personnel assisting the Coalition efforts with modernization projects and aiding the populace have often been ineffective.[3] The mixture of lush trees, rugged mountains, cross-border sanctuaries, and distrustful mountain residents have all contributed to creating a volatile mixture making the Korengal Valley a breeding ground for the insurgency.[4]

Task Force *Duke*

Soldiers from the 3d Brigade Combat Team, 1st Infantry Division, known as Task Force *Duke*, under the command of Colonel John M. Spiszer, deployed to the northeastern portion of Afghanistan along the Pakistan border from July 2008 to June 2009. Their area of operation specifically included Nangarhar, Nuristan, Kunar, and Laghman provinces. Executing Spiszer's counterinsurgency (COIN) strategy within their designated locations, TF *Duke* attempted to expand security and develop and train Afghan security forces while simultaneously enhancing the governance and economic expansion within their designated regions.[5]

Commanded by Captain James C. Howell, Bravo Company of 1st Battalion, 26th Infantry (1-26 IN) and hereafter known as *Viper* Company, attempted to attain Colonel Spiszer's COIN mission in the Korengal Valley.[6] Realistically, *Viper* Company's goal was "To keep the enemy tied down in the Korengal so that our battalion could make progress in the Pech River Valley."[7] This was anything but an easy task for roughly 150 Soldiers sparsely dispersed among several Combat Outposts (COPs), Firebases (FB), and Observation Posts (OPs) throughout the Korengal Valley. COPs and FBs differed mainly by the fact that combat outposts usually included the larger 120mm mortars while firebases might feature 82mm or 60mm mortars. *Viper* Company Headquarters (HQ) and the majority of 2d Platoon, a squad from the Battalion Scout Platoon, a 120mm Mortar section, Afghanistan National Army (ANA) Company HQ, and two ANA Rifle Platoons were located at the Korengal Outpost (KOP) nearly two kilometers west of Kandlay. A squad from 2d Platoon was located at OP Dallas, south of Aliabad. On the eastern side of the valley, 3d Platoon occupied COP Vegas and OP Rock, both less than a kilometer northeast of the village of Kandlay. Members of 1st Platoon were dispersed amongst three different locations: A fire team at OP 1 about a half kilometer west of the KOP. A quick reaction force (QRF) comprised of six men and two vehicles at the KOP and the remainder were assigned to FB Restrepo, northwest of Aliabad (see Figure 1 for map references).[8]

Viper Company's objectives were complicated by poor weather conditions that often plagued the valley and consequently limited available air assets. Convoys generally delivered supplies monthly to the KOP,

Firebase Vimoto (occupied by an ANA Rifle Platoon), and OP Dallas, the only three bases accessible by ground vehicles. Convoy replenishment required a "company-level effort" and Apache helicopters to ensure safe passage down Route Victory which was the only passable road through the valley.[9] On many occasions, the road was laced with mines and exposed to insurgent ambushes. Without the necessary supplies, especially ammunition, *Viper* Company sometimes discontinued combat patrols and remained in a defensive posture within their assigned outposts. Sometimes supplies were dropped by parachute requiring Herculean efforts to locate, gather, and protect the desperately needed materials and getting it safely back to the awaiting platoons. Occasionally, the cargo was dropped erratically into impenetrable sections of the valley or into the hands of the insurgents. Despite these imposing challenges, Captain Howell was determined to carry out his mission.

Enemy forces traveled freely throughout the valley allowing them to appear as normal persons, stashing caches of weaponry and supplies in designated areas. Many of these insurgents were born and raised in the Korengal Valley which only enhanced their ability to maneuver and attack incessantly. Firefights with the enemy occurred almost daily. "I've never understood how they can move in and set up on us without us knowing they were there," said Sergeant Craig Tanner, 3d Squad Leader.[10] Limited intelligence resources restricted *Viper* Company's ability to identify the locations of weapons caches and detained insurgents offered only sketchy details. However, *Viper* Company knew that the insurgents routinely used the same routes of travel.[11]

By sending out routine patrols, attempting to conduct weekly and monthly ambushes, and visiting neighboring villages, Captain Howell hoped to disrupt the elusive enemy and to gain some valuable intelligence on its operations in the valley. However, this task proved more difficult than initially expected. There were an inadequate number of *Viper* Company Soldiers to search the entire valley for weapons caches and high-value targets (HVT). Typically, the patrolling squads were unable to ambush insurgent forces since their route of travel was carefully observed by a "spotter," who notified enemy personnel over hand-held radio communication devices about the presence of US patrols.[12] However, this situation changed on 10 April 2009 when Second Lieutenant Justin R. Smith, leading his first combat patrol in Afghanistan, wreaked havoc upon the enemy.

Movement up the Ridge

For the mission that day, the Sawtalo Sar Ridge located approximately two and a half kilometers east of the KOP was chosen for 2d Platoon's objective because it was a favorite enemy site for routine attacks using DShK heavy and PKM general-purpose machine guns, rocket propelled grenades (RPGs), and AGS-17 automatic grenade launchers on the nearby Coalition outposts within the valley. Taking fire from multiple locations, *Viper* Company personnel assumed that there was only a team or squad sized insurgent element in the vicinity.[13] Captain Howell anticipated that by sending Lieutenant Smith on a clearing operation of 20 hours duration to the top of the ridge, he could identify the location from which the enemy fired these weapons, search for weapons caches, and disrupt any enemy operations throughout the night. Captain Howell neither gave Lieutenant Smith a specific location to establish a patrol base nor expected him to ambush the enemy. However, because no friendly forces had been up on this ridge in quite some time, Howell planned on having 2d Platoon conduct a lengthy patrol along Sawtalo Sar Ridge for half of the designated period and patrol back to the west for the other half. The plan included an overnight stay at a patrol base and, at first light the following day, a return to the KOP.

As a new *Viper* Company replacement officer with barely a week as 2d Platoon Leader, Lieutenant Smith, an honor graduate of the Army Ranger School, took his first assignment in Afghanistan quite seriously. Not wanting to disappoint either his platoon or his commanding officer, he prepared extensively for this mission. First, he studied several books pertaining to tactical instruction for Soldiers engaging in counterinsurgency operations including *The Bear Went over the Mountain: Soviet Combat Tactics in Afghanistan*. He also carried the *Ranger Handbook* which emphasized the five principles of patrolling: planning, reconnaissance, security, control, and common-sense tactics.[14] Second, Smith visualized the rugged terrain by standing on a bunker at the KOP where he scanned and planned his avenue of approach up the steep mountainside. Four days prior to the assigned mission, First Lieutenant John Rodriguez, Executive Officer for *Viper* Company, accompanied 2d Platoon on a reconnaissance patrol to familiarize Lieutenant Smith with the area and route of travel. This endeavor enabled Lieutenant Smith the opportunity to determine the estimated time it would take his platoon to reach the patrol base position, conduct map checks, and identify safe locations for water breaks. Then Smith factored in the anticipated time 2d Platoon needed for returning to the KOP. Third, Smith had a long discussion with his Non-Commissioned

Officers (NCOs) and discussed the mission itinerary, the social dynamics of the local villagers, and potential threats in the area. At this time, Smith was pleased to learn that his men were familiar with the *Ranger Handbook* since they were required to use it for their training.[15] Fourth, he rehearsed a number of "what if" scenarios repeatedly with his platoon. These included potential attacks by the enemy before crossing the Korengal stream and after, actions if sub-elements were separated, defense of the patrol base, and how the unit would employ its limited air assets.[16] Fifth, the platoon members conducted "rock drills" which provided them with the position and actions of each man throughout the duration of the patrol and placed particular emphasis on how to occupy the patrol base.[17] Because the company had not established a triangular patrol base since leaving the United States for its deployment, it was imperative for Smith to explain how he planned to set it up. Last, Smith continually consulted with his senior leadership to ensure that there was nothing that he had overlooked.

After extensive preparation, the 31 Soldiers of 2d Platoon headed out on the assigned mission. Included were six men from platoon's headquarters, nine from 2d Squad, eight from 3d Squad, and eight *Viper* Company Scouts assisting the platoon with its latest mission. On this patrol, 2d Platoon had a vast array of weaponry including three Mark 48 7.62mm light machine guns (a lighter version of the M-240), three Mark 46 5.56mm squad automatic weapons (SAW) which is a lighter version of the M-249, five 40mm grenade launchers (M-203s), an M-110 sniper rifle, one M-14 rifle, and 20 M-4 personal assault rifles. The lighter-weight weapons were test models given to *Viper* Company from the Asymmetric Warfare Group (AWG). Each Soldier wore his personal body armor, carried his assigned weapons with a combat load of ammunition (700 rounds for the M-4), transported sizable quantities of water, carried two to four grenades each, and hauled a wide variety of other gear required for the mission. This included personal radios, shovels, rations, etc.

They departed the KOP at 1130 with 2d Squad leading the way and 3d Squad bringing up the rear of the column. They began their lengthy trek down the "Stairway to Heaven" (villager carved steps into the mountainside in an attempt to ease travel) and crossed over the Korengal stream to the Kandlay Trail (see Figure 2) where they began an ascent up the steep ridgeline. Meanwhile, they stopped to sweep for weapons caches every 100 meters or 20 to 30 minutes walking distance apart. This operation required some Soldiers to take a "knee and post" security while others commenced their search. The procedure also allowed the men to take a much needed break, evaluate their surroundings for threats, and reduce the likelihood of heat exhaustion as they ascended up the steep ridgeline. Using spotting scopes, *Viper* Company Soldiers back at the KOP were able to monitor the progress of 2d Platoon as they took their breaks and kept an eye out for enemy observers and other potential threats.[18]

Throughout this journey, Lieutenant Smith and his men were routinely observed by local villagers who watched and reported every move to the insurgency. Even though this was a common occurrence in the valley, this time it favored 2d Platoon. For unknown reasons, the enemy spotters lost track of the Soldiers during their ascent up the ridge and reportedly assumed that they had already turned back towards the KOP. This constituted a fatal mistake. Prior to this patrol, *Viper* Company normally established their ambushes and patrol bases during daytime hours. Departing before dawn and setting up their ambush before the sun rose, the Soldiers waited throughout the day in hopes of spotting the enemy. If no enemy appeared before nightfall, they packed it up and departed. This time though, they mixed things up a bit and remained on the Sawtalo Sar Ridge throughout the night. This deviation from the standard operating procedure (SOP) greatly contributed to 2d Platoon's successful mission.[19]

While walking along the Kandlay Trail, the Soldiers discovered a smaller trail that intersected with the nearby Chichal/Donga Trail (see Figure 3) which ran north to south across the Sawtalo Sar Ridgeline. It was on this spur that 2d Platoon noticed signs of personnel traffic through the area such as food remnants, discarded batteries, prints from varying footwear, and empty water containers dropped haphazardly along the trail. Lieutenant Smith radioed Captain Howell informing him of their findings and confirming that no other friendly units had been in the area. After alleviating his concerns, Smith decided to modify his pre-planned patrol base location near a landing zone on top of Sawtalo Sar. He believed that the enemy crossed from one side of the valley to another at this point. With this recent discovery, Smith opted for a better position to monitor this portion of the heavily used trail. After marching uphill for six and a half to seven hours, 2d Platoon finally reached the objective rally point (ORP), roughly 1900 meters in elevation and two and a half kilometers away from the KOP.[20]

Figure 2. 2nd Platoon Soldiers on the Kandlay Trail.

Figure 3. The Chichal-Donga Trail looking north. (Left to Right) Staff Sergeant Little, Lieutenant Smith, Sergeant Reese, and Sergeant Nightingale.

NOTE: Sergeants Reese and Nightingale are wearing the Plate Carrier type ballistic vests with their Modular Lightweight Load-carrying Equipment (MOLLE) gear while Staff Sergeant Little and Lieutenant Smith are wearing the heavier and more restrictive Improved Outer Tactical Vest (IOTV). At the time of this photo on 10 April 2009, the Plate Carriers were not authorized but completely essential for the work of a Scout/Sniper. The Plate Carriers are now standard issue for those military personnel in Afghanistan not typically riding in vehicles.[21]

By this time, the sun had nearly set. The platoon was approximately 100 meters from the new patrol base location. Smith led a patrol that included a scout and three Mark 48 gun teams comprised of a gunner and his assistant to establish the patrol base by placing the gun teams in the apexes, 9, 3, and 6 o'clock positions of his triangular patrol base. Once these were set in place, the remaining Soldiers took their positions. The 2d Squad situated itself southeast looking towards the trail and the enemy's approach. The bulk of 3d Squad was positioned on the southwest side of the patrol base facing a ravine. The majority of the Scouts faced northwest and conducted rear security within the patrol base. Each Soldier knew who was on his left and who was on his right prior to settling in. Once in place, some men began digging in while others could only modify their gun positions because parts of the ridge were solid rock.

As the men prepared their fighting positions, squad and team leaders established firing sectors and further improved their fighting positions.[22] Three claymore anti-personnel mines were concealed in strategic locations. One was positioned near the 6 o'clock apex designed to cover the backside of the patrol base. Sergeant Thomas Horvath and Private First Class Richard Dewater positioned a second claymore facing southeast to cover a portion of the trail. The third mine covered the only remaining route for the enemy's escape to the north. This claymore was manned by two Scouts, Specialists Joseph Raynal and Matthew Spencer, who were also providing rear security for the patrol base.

At this time, Lieutenant Smith led a scout recon team up the Sawtalo Sar Ridge to identify a favorable place for setting up the observation post/listening post (OP/LP). Smith determined that the OP/LP would be positioned nearly 75 meters from the top of the ridge and 80 to 100 meters from the patrol base. Several Scouts, including the squad leader Staff Sergeant Christopher Little, senior sniper and team leaders Sergeant Philip Nightingale and Sergeant Zachary Reese, and the grenadier Specialist Jordan Custer joined Lieutenant Smith in this endeavor. In addition, communication specialist Private First Class Stuart Healey from the company headquarters joined the recon team and assisted in collecting

Global Positioning System (GPS) grid positions every 20 meters along the trail. This provided precise trail coordinates in latitude and longitude for use as targeting reference points for incoming artillery and mortar rounds.[23]

Around 1830 hours when dusk turned into night, the Soldiers went on "stand-to" alert, a period when all personnel are awake and focused on their assigned sectors watching and listening for any signs of movement. Using the cover of the night and setting up the patrol base in a bushy well-concealed area, 2d Platoon's position was hidden from insurgents using the trail. An hour later, Lieutenant Smith sent Sergeants Reese and Nightingale and Specialist Custer to the OP/LP position southeast of the patrol base and west of the newly discovered trail. Carrying their equipment bags with them, the Scouts walked the short distance in approximately three minutes. Originally, they planned on digging in and constructing a place to hide so that they could monitor the trail but the approaching enemy interrupted this plan.[24]

Ambush

Shortly after arriving at the OP/LP, Sergeant Reese noticed movement on the trail. At first, Reese thought the shadowy figures were goats but immediately realized his error. He radioed Lieutenant Smith and advised him of a "large group of armed males" walking down the trail from the east. The insurgents had come over the top of the ridgeline and followed the Chichal-Donga trail to the north.[25] Due to the darkness, Reese was unsure of the exact number of insurgents, until the second or third man in the column switched on and off a small LED-type flashlight, allowing Reese to initially count eight individuals at the front of the column.[26] As the insurgents headed toward the patrol base, Staff Sergeant Little upon hearing Sergeant Reese's radio transmission, repositioned himself to better observe the enemy heading his way. At first Lieutenant Smith thought something was wrong when Sergeant Reese radioed him as it had been less than 15 minutes since the OP/LP was emplaced. In the meantime, the OP/LP trio went prone and hid behind a felled tree. They were there to observe and only shoot when absolutely necessary or upon command. Smith, once told the enemy's location, pulled Soldiers from other positions to reinforce the new kill zone south to southeast of the triangle between the 9 and 3 o'clock positions. The patrol base had now become a modified linear ambush position.

After counting 26 insurgents within three meters of the OP/LP site, Sergeant Reese was unable to communicate the enemy's increased numbers to Lieutenant Smith without jeopardizing his position. Once the enemy came into view, Smith was perplexed. There were three or four spread out but a large group of them were all bunched together, strolling down the trail as if they just departed a *shura* in the local village. "I'm like, oh man, what is going on? Are the villagers lost up here?" Smith wondered.[27] He was reminded of a recent statement from Captain Howell, "There's nobody who's going up there at that time of night for anything good."[28] The Platoon Leader scanned the group of men for signs of weapons. The entire US platoon was wearing night-vision equipment providing them a significant advantage over the enemy. Sergeant Little used an AN/PEQ-15 Advanced Target Pointer Illuminator Aiming Light (ATPIAL) infrared device which provides a beam of infrared energy that helps illuminate targets. Lieutenant Smith was able to positively identify 15 men congregated within a 40-meter section of the trail carrying weapons and wearing ammo pouches entering the 50-meter wide kill zone.[29] In this zone, Smith had maximized his firepower with a claymore mine, four machine guns (two Mark 48s and two Mark 46s), an extra M-203, and other personal weapons.

By this time, Sergeant Reese's warning, "Enemy personnel approaching," had already been transmitted or passed down the line. Many of the Soldiers had never seen the enemy this close and in such great numbers. Private First Class Brad Larson, 2d Squad grenadier, recalled, "My heart started pounding and I was scared shitless just waiting for them to get in sight."[30] As the men of 2d Platoon mentally and physically prepared for the impending attack, they trained their weapons on the insurgents coming into view. Sergeant Joshua Trudel, 2d Squad Leader, could hear the enemy talking and prepared to fire his weapon.[31] The impatient but trained men were awaiting the command to open fire but because of the trees that obstructed his view, Smith still could not see the end of the enemy column.

The enemy walked down the crest of the Sawtalo Sar Ridgeline following the Chichal/Donga Trail northwest until it opened up into a small clearing. At this point, the trail doglegged to the right changing to a southwesterly direction. This meant the insurgents were moving into the kill zone at an acute angle crossing from left to right (see Figure 4).

As Lieutenant Smith readied to trip the claymore, the nearest machine gunner to his right "hit the switch on his Mark 48," making a metallic clicking sound.[32] Sergeant Trudel said the enemy point man stopped about six feet away upon hearing the noise and he had a "confused look on his

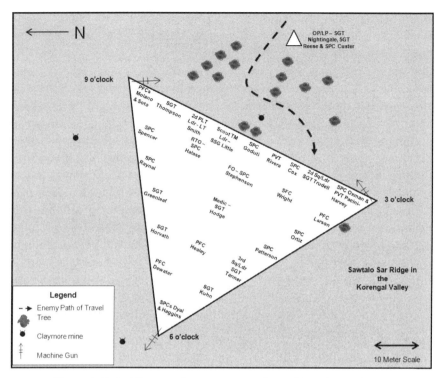

Figure 4. Patrol Base and enemy route of travel into the ambush.

face."[33] The insurgents began bunching up when the lead man heard the noise and looked toward Smith's position. Fearing that the element of surprise had been lost, the lieutenant gave the command to open fire. Approximately 15 insurgents were within the kill zone many of which were only three to six feet away from 3d Squad machine gunner Specialist Robert Oxman who was in the 3 o'clock apex position of the triangle along with his assistant gunner, Private Troy Pacini-Harvey.

"Within seconds of the last person walking by our OP/LP, the patrol base initiated contact," said Sergeant Reese.[34] Lieutenant Smith whispered over the radio, "Fire, fire, fire."[35] Four of the enemies were directly in front of Trudel's location. At this point, every Soldier facing the trail opened up on the insurgents who attempted to fight back, died trying, or ran away. Smith blew one claymore sending shards of shrapnel into the enemy column. Private First Class Larson, nearest to the southern ravine, noticed two enemy heading in his direction but their close proximity prevented him from firing his 40mm grenades which require a minimum of 15 meters to arm. So he used his rifle instead. One fell within six feet of his position and Larson remembered, "The other fell and rolled down the cliff."[36]

When the wounded insurgent at Larson's feet attempted to reach for his weapon, Larson killed him with a shot to the head. Sergeant Tanner saw three or four insurgents from the back of the enemy column head toward a draw that contained a claymore and yelled to his men to blow the mine and throw grenades. Sergeant Christopher Thompson, A Team Leader for 2d Squad, launched a 40mm grenade from his M-203 on a group of the enemy hitting three to five of them. Lieutenant Smith saw an arm flying off one of them. Private First Class Arturo Molano, a machine gunner for 2d Squad, hit one fleeing insurgent as he fell down into a ravine (see Figure 5).

When Captain Howell and Lieutenant Rodriguez at the KOP received the first reports from Specialist Steven Halase, 2d Platoon's radio telephone operator (RTO), they were unsure who was ambushing who. As the firefight dwindled, more details filtered into *Viper* Company HQ. Soldiers at the KOP saw tracer rounds piercing the night sky at the ambush location. Simultaneously, 3d Platoon Soldiers at OP Rock and COP Vegas witnessed flashlights frantically moving eastward over the ridge towards the village of Chichal. After action reports suggest that these were the 12 enemy personnel who had not yet reached the kill zone when Lieutenant Smith's men opened fire. However, once 2d Platoon's location was determined, as a safety precaution 3d Platoon opened fire on the escaping insurgents with their M-240s nearly a kilometer away.

The enemy, caught totally by surprise, continued running in all directions during the deadly onslaught. In fact after the cease fire, platoon members discovered a right shoe and a left shoe about four meters apart with a blood trail leading towards a cliff edge. Apparently, one insurgent was either blown out of his shoes or ran right off a cliff. [37] At the OP/LP, the Scouts remained prone as friendly fire flew over their heads. As soon as they noticed several insurgents running in their direction, they too joined in the fray. Sergeant Nightingale shot two of the enemy with his silenced M-4 potentially saving Sergeant Reese from harm because the enemy was within a few feet of Reese's location. Nightingale then threw a grenade at another insurgent heading south down the trail. Specialist Custer used both his M-4 and M-203 to fire on the fleeing men. Reese shot another insurgent with his M-110 Sniper rifle and threw a hand grenade in his direction.

After approximately 10 minutes, the men of 2d Platoon ceased fire and began looking for enemy survivors, collecting the dead, and conducting a sensitive site exploitation (SSE) by searching the dead insurgents. [38] Platoon members also checked to see if any of their team had been

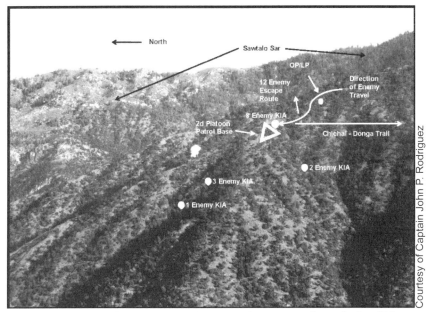

Figure 5. Location of patrol base and enemy elements on Sawtalo Sar Ridge.

wounded or killed and reconsolidated their ammo. Aiding in this search, the KOP fired 120mm illumination mortar rounds turning night into day. Private First Class Larson found a box of animal crackers, some hashish, and typical military paraphernalia on one body. Sergeant Trudell, who had his camera with him, took photographs of the bodies for intelligence purposes while his men searched and cataloged items from four dead insurgents in front of their position.[39] Two or three of the enemy were obviously hit by the shrapnel from a claymore mine due to the peppering of metal in their faces and riddled holes in the AK-47 automatic rifles.[40]

The Scouts began their SSE near their OP/LP as well. Sergeant Nightingale moved two bodies to the collection point while Sergeant Reese and Specialist Custer provided security. After Nightingale returned, Reese began cutting away the rack system (ammo pouches) from another insurgent who appeared to be dead. He wasn't. The wounded insurgent suddenly grabbed Reese. The sergeant then swiftly plunged his knife into the man's eye socket. Platoon HQ directed the three Scouts to return to the patrol base because air support was arriving. Lieutenant Smith needed to ensure that all of his men were accounted for prior to the use of any air assets on the ridge. Running back to their base, the Scouts shouted the password "Budweiser, Budweiser," to prevent being shot at by other members of the platoon.[41]

Captain Howell had called in a pair of F-15 Strike Eagle aircraft for combat air support (CAS) as soon as the ambush started and they arrived approximately 20 minutes later at 1945 hours. Soldiers from COP Vegas noticed flashlights from suspected insurgents maneuvering through a deep ravine just south of Chichal heading toward the ambush site, an apparent attempt to assist their ambushed comrades nearly one kilometer away. Howell directed an F-15 fighter to drop a 2,000 pound bomb on the alleged enemy reinforcements. Members of 3d Platoon sent out a patrol to conduct a bomb damage assessment but were unable to get to the bomb site at the bottom of a creek bed.[42]

For the next hour or so, the surviving insurgents attempted to evade the platoon. An Air Weapons Team (AWT) comprised of two Apache AH-64 helicopters arrived and joined the search north and south of the patrol base. Lieutenant Smith remembers that, "There were still a couple of bad guys out there shouting to each other *delta rasha* or 'come to me'" and he used this to his advantage when trying to find them.[43] Whenever the enemy voices were heard, the Soldiers fired their weapons and threw grenades in the direction of the sounds in hopes of eliminating potential threats. After two or three incidences of this, the insurgents remained silent.

The Apaches equipped with thermal imagers assisted in the nighttime search. Several insurgents attempted to hide in a small cave on the north side of the cliff. Smith requested an Apache to make a 30mm gun run on this group because the infantrymen could not readily locate them. The helicopter succeeded in killing three insurgents. Captain Howell believed that a HVT might have been within the enemy column and ordered the 2d Platoon to collect the bodies. Two Scouts, Staff Sergeant Little, and Sergeant Reese along with Forward Observer Specialist Kyle Stephenson accompanied Lieutenant Smith who attempted to maneuver down the precarious ridge. While sliding down the treacherous cliff however, Reese lost his footing on a steep slope and tumbled nearly 50 meters before hitting his head knocking himself unconscious. Smith, fearing that Reese was dead, joined Stephenson in a frantic climb down to the fallen Soldier's body. Cradling Reese in his arms, Smith asserted, "Reese, wake up, Reese, wake up." Finally Reese awoke in puzzlement, looked up at the lieutenant and queried, "Why are you holding me?" Relieved that Reese was okay, Smith ordered the sergeant to take a few moments to collect his senses.[44]

Around 2100 hours, *Viper* Company Soldiers from FB Restrepo using thermal imaging devices identified a lone insurgent attempting to rendezvous with his imperiled cohorts on the ridge. He apparently departed the village of Donga and was heading up the Donga Trail towards the ambush site when he was spotted. Lieutenant Rodriguez at the KOP called in the Apaches. Using a long-range infrared laser pointer/illuminator, an IZLID, he guided the Apaches to the insurgent who was crouched down near a rock along the trail. The Apaches opened fire with 2.75mm rockets and 30mm rounds. Some 3d Platoon Soldiers at FB Restrepo reported that the insurgent did not appear to be moving. Not sure if the man was dead or just pretending, the Apaches were ordered to make another gun run on the target blowing the insurgent to pieces.[45]

Ultimately, the ambush killed 15 enemy personnel. The 2d Platoon recovered 13 bodies and two more were confirmed at the bottom of a steep ravine. The platoon also confiscated 10 AK-47 assault rifles, one RPG launcher, two RPG rounds, five ammo pouches, 30 ammo magazines, and two grenades. Initially, three RPG rounds were found but one was considered as too unstable for transporting back to the KOP and was destroyed at the scene. The US Soldiers also discovered five hand-held radios, one cell phone, two cassette tapes, one radio, one calculator, religious materials, Pakistan currency, flashlights, and mirrors. Many of the insurgents were wearing battle dress uniforms (BDU) and black military boots. Others were wearing the traditional *shalwar kameez* "man jammies" with tennis shoes or sandals.[46]

No Soldiers were wounded or killed during the engagement. Only two members of the platoon had minor injuries. Sergeant Reese received a concussion from his fall and Private First Class Larson twisted his ankle sliding down another steep ridge. Due to the precipitous terrain, the long walk back to the KOP, and the unsteadiness of the two injured men, Lieutenant Smith had Reese and Larson evacuated by a UH-60 Black Hawk helicopter to Jalalabad Air Base for medical treatment.[47]

Once the ambush occurred, 2d Platoon's original mission was compromised and there was no point in staying in the area. So the remaining members of the platoon departed the ambush site around 0230 hours and began their arduous journey back to the KOP. In an attempt to avoid any possibility of retaliation, Apache helicopters flew overhead watching the area with their thermal imagers. Moving down a mountain during daytime was strenuous enough but 2d Platoon took on the task in absolute darkness. They arrived at the KOP 10 to 12 hours later, both mentally and physically exhausted. Captain Howell welcomed them back by having the cooks prepare hot food and congratulating them on a job well done.[48]

The next morning, elders from the villages where the dead insurgents lived, requested a *shura* with Howell at the KOP. The elders were told of the ambush and given permission to collect their dead from the Sawtalo Sar Ridge. The elders claimed that the men who died were nothing more than a rescue team searching for a kidnapped village girl. Sergeant Reese recollected Howell's response to the elders, "That is outrageous! Who takes RPGs and hand grenades to rescue a little girl?"[49] The men of 2d Platoon watched from the KOP as the local villagers headed to the ambush site to collect their dead. Using the dead men's beds as litters, the villagers wrapped the bodies in cloth and carried them back for a funeral later that day. As they descended along the Kandlay Trail, the locals split off to nearby villages. All of the dead insurgents were from neighboring villages in the Korengal Valley. The men of *Viper* Company witnessed the funerals from the KOP but had no remorse for the deceased.[50]

Aftermath

The success of 2d Platoon's mission on the ridge was based on several critical factors. The first of these was the leadership of Second Lieutenant Smith who personally planned and conducted pre-mission rehearsals, ensuring that each Soldier knew his position and duties automatically. Second was the change in operational pattern that led to the enemy observer's assumptions about the platoon returning to the KOP. This shift was vital to *Viper* Company's success in launching the ambush. Third was the timing of the platoon's movement. If the platoon had been an hour or so later in arriving, there is a good chance that they would have walked into an ambush or at least found themselves in a lengthy firefight with the enemy. Fourth was the flexibility displayed by the men of 2d Platoon. Initially, 2d Platoon planned on establishing a patrol base that would allow them to monitor the trail. However, the unexpected presence of the enemy insurgents forced the platoon to rapidly convert their patrol base to a modified linear ambush position. Sergeant Reese said, "Operations can change at a moment's notice for any number of reasons…you will stay alive by keeping it simple and using your common sense."[51] Fifth, was the operational discipline with which 2d Platoon operated. "If it wasn't for them [2d Platoon Soldiers] having the discipline that they had and trust within each other, then things could have turned out a lot different that night," said Smith.[52] This last factor was perhaps the most critical given the historical importance of noise and light discipline in small unit ambushes.

Observers may note that the ambush on Sawtalo Sar Ridge did not rigidly follow Army templates for ambush operations. According to the *Ranger Handbook* and other Infantry manuals, flank security should be posted on either side of an ambush position. Second Lieutenant Smith did not have the manpower to do this. He had chosen to beef up the ambush line and man other crucial positions instead. The ambush on the ridge was less than perfect in other ways as well. After all, 12 insurgents did manage to escape the trap but after the operation, Smith noted that the terrain restrictions limited the length of the kill zone. Further, 2d Platoon had an insufficient numbers of Soldiers to spread further along the trail without compromising safety and security. The first 15 insurgents entered the kill zone at an angle when the last enemy element in the column had just passed the OP/LP nearly 80 to 100 meters away. This created a gap that was not covered by fire. According to Smith, he would have needed twice the number of Soldiers to cover the area sufficiently. In the end, however, the platoon leader's decisions and the professionalism of his Soldiers allowed for one of the most successful small-unit actions in the history of US operations in the Korengal Valley.

Five days after the ambush on 10 April 2009, Private First Class Dewater was killed by an improvised explosive device (IED) as he crossed the Korengal stream between the villages of Aliabad and Laniyal. The other members of 2d Platoon then came under a brutal barrage of small arms fire from multiple locations, forcing the Soldiers to take cover until a pair of A-10 Thunderbolt II ground attack aircraft were able to quell the heavy enemy fire. Captain Rodriguez and Lieutenant Smith noted that insurgent use of IEDs along trails and near villages was an uncommon occurrence and strongly believed these were acts of retaliation for the successful ambush on Sawtalo Sar Ridge.[53]

Notes

1. National Geospatial Intelligence Agency. *Afghanistan Country Atlas Series Volume 4: Badakhshan, Kabul, Kapisa, Kunar, Laghman, Logar, Nangarhar, Nuristan, Paktiya, Parwin, Scale 1:50,000.* National Geospatial Intelligence Agency, 2009.

2. C.J. Chivers. "Pinned Down, a Sprint to Escape Taliban Zone." *New York Times,* 20 April 2009. Info Note – The 35 KIAs depict those that actually died in the Korengal Valley from hostile fire according to the Military Casualty Information website. Media sources annotate higher numbers of deaths ranging from 40-42 Soldiers. However, these include non-hostile fire deaths like (vehicle accidents, health, & suicide, etc.) and some are inaccurately identified to specific villages, mountain tops, and military bases within the Korengal Valley.

3. Michael Moore and James Fussell. *Afghanistan Report 1: Kunar and Nuristan: Rethinking US Counterinsurgency Operations.* (Washington DC: Institute for the Study of War, July 2009), 7; Elizabeth Rubin. "Battle Company is Out There." *New York Times Magazine,* 24 February 2008, 1-2.

4. Williams, Brian Glyn. "Afghanistan's Heart of Darkness: Fighting the Taliban in Kunar Province." *Combating Terrorism Center at West Point - Sentinel* 1, no. 12 (November 2008): 11.

5. Colonel John M. Spiszer. Teleconference from Afghanistan by Colonel Gary Keck, 18 November 2008. News Transcript. US Department of Defense, Washington, DC, 3.

6. Colnel John M. Spiszer, interview by Steven E. Clay, Combat Studies Institute, Fort Leavenworth, KS, 22 November 2010, 2.

7. Captain John P. Rodriguez, interview by Scott J. Gaitley, Combat Studies Institute, Fort Leavenworth, KS, 24 May 2011, 3.

8. Rodriguez, interview, 10; Captain John P. Rodriguez. Journal Entries 5 – 6 April 2009. Copied and sent to Scott J. Gaitley.

9. Captain John P. Rodriguez, interview by Scott J. Gaitley, Combat Studies Institute, Fort Leavenworth, KS, 24 May 2011, 3.

10. Sergeant Craig W Tanner, interview by Scott J. Gaitley, Combat Studies Institute, Fort Leavenworth, KS, 5 July 2011, 13.

11. Major James C. Howell, e-mail to Scott J. Gaitley, 14 April 2011, Combat studies Institute, Fort Leavenworth, KS.

12. Bing West. *The Wrong War: Grit, Strategy, and the Way Out of Afghanistan.* (New York: Random House, 2011), 34.

13. Rodriguez, interview, 9.

14. Department of the Army. *SH 21-76, Ranger Handbook.* (Fort Benning, GA: Ranger Training Brigade, July 2006), 5-1.

15. Tanner, interview, 12.

16. Captain Justin R. Smith, interview by Scott J. Gaitley, Combat Studies Institute, Fort Leavenworth, KS, 6 July 2011, 5.

17. Smith, interview, 4-5; Tanner, interview, 4.

18. Rodriguez, interview, 12.

19. Rodriguez, interview, 12.

20. Rodriguez, interview, 10; Smith, interview, 6.

21. Zachary Reese, e-mail to Scott J. Gaitley, 3 August 2011, Combat Studies Institute, Fort Leavenworth, KS, 1.

22. Department of the Army, *Field Manual 3-21.8, The Infantry Rifle Platoon and Squad.* (Washington, DC: Department of the Army, 2007), 8-20; Smith, interview, 9.

23. Reese, e-mail, 6 July 2011, 1; Tanner, interview, 9.

24. Reese, e-mail, 6 July 2011, 1.

25. Reese, e-mail, 7 July 2011, 1.

26. Reese, e-mail, 6 July 2011, 1.

27. Smith, interview, 11.

28. Smith, interview, 11.

29. Smith, interview, 14.

30. Brad Larson, e-mail to Scott J. Gaitley, 7 July 2011, Combat studies Institute, Fort Leavenworth, KS, 2.

31. Staff Sergeant Joshua R. Trudel, e-mail to author, 14 July 2011, Combat Studies Institute, Fort Leavenworth, KS, 1.

32. Smith, interview, 14.

33. Trudel, e-mail, 1.

34. Reese, e-mail, 6 July 2011, 1.

35. Tanner, interview, 5.

36. Larson, e-mail, 2.

37. Smith, interview, 19.

38. US Department of the Army. *Field Manual 3-90.15*, *Site Exploitation Operations.* (Washington, DC: Department of the Army, 2010), 1-1.

39. Trudel, e-mail, 2.

40. Smith, interview, 12.

41. Reese, e-mail, 6 July 2011, 2.

42. Exhibit 1 (2/B Storyboard), "Sawtalo Sar Ridgeline Patrol," 10 April 2009, 2; Rodriguez, interview, 16.

43. Smith, interview, 14.

44. Reese, e-mail, 6 July 2011, 2; Smith, interview, 18.

45. Exhibit 1 (2/B Storyboard), "Sawtalo Sar Ridgeline Patrol," 10 April 2009, 3; Rodriguez, interview, 17; Rodriguez. Journal Entries 10 April 2009, 1.

46. Smith, interview, 19.

47. Larson, e-mail, 2.

48. Trudel, e-mail, 2.

49. Reese, e-mail, 3 August 2011, 1.

50. Exhibit 1 (2/B Storyboard), "Sawtalo Sar Ridgeline Patrol," 10 April 2009, 3; Rodriguez, interview, 18; Tanner, interview, 16.

51. Reese, e-mail, 6 July 2011, 2.

52. Smith, interview, 20.

53. Smith, interview, 30; Rodriguez, interview, 30.

Flipping the Switch
Weapons Platoon Movement to Contact in Zhari District
by
Michael J. Doidge

Background

Speaking before a class of West Point cadets in December of 2009, President Barack Obama announced his plan to dramatically increase US forces in Afghanistan. Known informally as "the surge," the influx of military manpower palpably represented the strategic reorientation of US military efforts, a signal that the United States was placing Afghanistan back to the forefront of its war on terror.

Likening the war to a fight against cancer, President Obama, in coordination with his military advisors, sought to gradually introduce the surge forces throughout Afghanistan to fight the Taliban's spreading influence in key locations. For its part in the surge, the US Army's 2d Brigade Combat Team (2BCT), 101st Airborne Division drew the Zhari district as its area of operation. Situated in Kandahar province just west of Kandahar City, Zhari was a *mujahedeen* stronghold during the Soviet occupation in the 1980s. During the 1990s, the Taliban protected it with successive belts of improvised explosive devices (IEDs) as well as a nebulous and far-reaching shadow government that enjoyed complete subservience from the local populace. The Soviets' nickname for Zhari was appropriate: "The Heart of Darkness." The three battalions comprising 2BCT, the 1st Battalion, 502d Infantry Regiment (1-502 IN), the 2d Battalion, 502d Infantry Regiment (2-502 IN) and the 1st Squadron, 75th Cavalry Regiment (1-75 CAV), were tasked with shattering the Taliban's hold on the area, preventing future incursions from Pakistan, and clearing the southernmost area's "green zone" in order to reintroduce farmers to the land. Given the vast mission objectives and limited time with which 2BCT had to work, the task before it was nothing short of monumental.

When the 1-502 IN arrived in theater in May of 2010, it assumed control from the 1st Battalion, 12th IN, 4th Infantry Division (4ID). The sizeable area of operations stretched the 1-12 IN thin. Now however, the surge was about to relieve a battalion with a brigade. This expanded mission capabilities for US forces in Zhari and the 1-502 IN's first month in theater reflected this change accordingly. Tasked with developing situational awareness for the rest of the brigade and expanding upon previous tactical infrastructure, of the three battalions, the 1-502 IN possessed the most population-centric mission in the brigade. Recognizing this, the battalion assigned Delta Company to collect information on the town of Senjaray. From there, the 1-502 IN's operations would shift south across Highway 1, a paved road that ran east-west through 2BCT's area of operations. Everything south of Highway 1 was under strict Taliban control. For 3d Platoon of Delta Company, 1-502 IN, the shift south began on 23 June 2010.

Flipping the Switch

From a bridged intersection just southwest of Senjaray on Old Highway 1, or Route RED STRIPE as it was called by the Soldiers of the 1-502 IN, local Taliban forces ran an extortion operation aimed at intimidating Senjaray's population of 10,000 into subservience. Whether the toll came in the form of money or goods mattered little, everyone paid. Delta Company witnessed it daily from Combat Outpost (COP) Senjaray. Located at COP Senjaray was a Joint Land Attack Cruise Missile Defense Elevated Netted Sensor (JLENS), an aerostat carrying a camera with precision zoom. The Soldiers of Delta Company used the camera to identify the Taliban, their weapons, and their shakedowns.[1] In terms of physical distance, the space between the JLENS and the intersection was easily measurable in a few short kilometers. Yet the camera may as well have been looking into another world. No US Soldiers had stepped foot near this Taliban stronghold since 2005 and the brazenness of Taliban efforts in the area reflected their belief that US troops would not be back anytime soon. If the battalion's counterinsurgency efforts in Senjaray held any hope of winning over popular support, they needed to end local extortion, return to the population the freedom to sell their goods with impunity, and eliminate the Taliban's iron grip over the local population.

The 1-502 IN planned to engage the Taliban through both lethal and non-lethal means. Before the battalion could conduct lethal operations, it needed situational awareness over the population in the small city of Senjaray. That meant conducting "white pages" missions. First Lieutenant Evan Peck of 3d Platoon referred to "white pages" as "a door-to-door assessment, kind of soft knock, where we try to gather the pictures of everybody there, get a grid to their door, and try and get an assessment of the population as to how many males and females live in each compound."[2] At 0500, 3d Platoon set out on 23 June to perform what was by now a daily ritual. Peck stated "White pages sends a good [information operations] message...It lets the people know that we care about them...that we recognize who lives in this community."[3] The battalion's information operation objective in Senjaray was to create a mutually shared stake in the community between US troops and the local population. The effort linked the people of Senjaray to friendly US forces tasked with building infrastructure, keeping the population safe, and improving quality of life. US forces hoped to reap results at the tactical level in the form of popular support for community projects that would erode the Taliban's influence.

By familiarizing the platoon with the most important aspects of Senjary's religious and social life, platoon members came to know and remember familiar faces, patterns of life, and how Senjaray's inhabitants made their living. Conducting repetitive white pages missions protected the people of Senjaray and the platoon by identifying abandoned compounds and singling out hostile persons. Examining the population of Senjaray's patterns of life also enabled battalion to root out the enemy within. The S2 battalion intelligence officer took 3d Platoon's findings and created spider webs of social networking data replete with information on city inhabitants who held legitimate jobs as well as those who made their living through unidentifiable means. Repetition was the key to successful white pages operations and 3d Platoon believed this day's mission would not be any different. Then an unexpected explosion south of Senjaray alerted the platoon that something big was going down.

The blast was the result of an A-10 strike called in by a US Special Operations Task Force whose purpose was to track high-value targets across Afghanistan's Southern Provinces. [4] Before the explosion, 1-502 IN Command Post (CP) had been alerted that an airstrike could be conducted in the area south of Senjaray and was told to stay out of the area to avoid alerting the enemy.[5] On 23 June 2010, suspected Taliban leaders were identified meeting in a known homemade explosive (HME) factory in the area. The A-10 attacked the building. Despite warning the 1-502 IN CP, Peck's Platoon was surprised by the explosion.

Immediately following the strike, the platoon rapidly moved south to Highway 1 to get eyes on the blast site. In so doing, 3d Platoon placed themselves closest of any US unit to the strike location. By virtue of their location and 3d Platoon's organic resources, the Delta Company Commander decided that 3d Platoon was capable of conducting a battle damage assessment (BDA) and ordered them to do so. Moving across Highway 1, Peck radioed the company CP and visited the local Afghanistan National Police (ANP) to see if they had any information on the strike. The platoon leader was on good terms with the police chief. Because of this, the chief assigned Peck four ANP policemen to travel with the platoon.

While Peck spoke with the ANP, Sergeant First Class Michael Calderaro, the Platoon Sergeant, fixed the location of the strike and developed an approach route to the site. Third Platoon also contacted Delta Company to coordinate reinforcements and fire support for their movement south. The Delta Company Commander designated 4th Platoon as the mission's Quick Reaction Force (QRF) and also arranged for two OH-58 Kiowa helicopters and one AH-64 Apache to provide close combat attack (CCA) support.

Third Platoon, Delta Company, was a weapons platoon comprising 16 Soldiers for its mission on 23 June. Peck's headquarters element included the platoon radio telephone operator (RTO), Private First Class Joshua Keck, and the unit medic, Private First Class Robert "Doc" Rankin. To that Peck added two interpreters and four ANP personnel. Sergeant Matthew Hetrick aided Calderaro, serving as an assistant platoon sergeant who would augment the HQ element in any combat scenario that might evolve, and shifting resources during the battle as necessary to adapt to conditions on the ground. As the platoon prepared to move, Hetrick and Calderaro conducted troop leading procedures, rehearsed the route, and ran pre-combat checks. Hetrick located himself with 3d Platoon's first squad led by Sergeant Daniel Bartlett while Calderaro located himself with the second led by Sergeant Nicholas Hays. The platoon was ready. They had "flipped the switch" as Hetrick later stated. "The 23rd of June started as a regular old, you know, knock on the peoples' door, be nice to them, patrol, and then there's an air strike and 30 minutes later each and every guy had his switch turned on."[6] At 0500 3d Platoon had set out to conduct a routine white pages mission. By 0900 they began their move south to become the first US unit to set foot on Route RED STRIPE since 2005.[7]

Once 3d Platoon moved south, they did so with the knowledge that they would not have indirect fire support. If 3d Platoon were cut off or surrounded by enemy forces, they could potentially call for fire from two sites: the 81-mm mortars at COP Senjaray under the control of A Company and the 120-mm mortars at COP Ashoque under control of Headquarters Company (HHC). However, Senjaray was a heavily populated area and the battalion wanted to limit use of mortar fire for fear of causing civilian casualties.[8]

Third Platoon set out south in wedge formation from the ANP station. As they moved through a nearby cemetery, it granted them freedom of movement. The platoon spread out to 10 to 15 meters between individuals and approximately 50 meters between squads. As Peck's platoon made its way south, it used simple but effective techniques designed to lessen the possibility of IED detonations. Those techniques included placing Soldiers who had previously served in Iraq and Afghanistan in the lead element to use their experience in identifying potential hazards. The majority of IEDs in Eastern Zhari were triggered by pressure plates placed on foot paths or chokepoints where people were forced to walk. Tapping into the enemy's psyche, Sergeant Hetrick argued, "The enemy took advantage of human nature and placed [IEDs] on paths of least resistance."[9] Wagering that Soldiers would naturally move through a break in a wall rather than climb over it, the enemy would place them in the walls near the gap. As a counter-measure, the platoon stayed away from footpaths, maintained spacing between Soldiers, and refused to move through choke points choosing instead to climb walls, walk uneven terrain, and generally pursue the more difficult routes of approach. The platoon walked through waterways and flooded canals whenever it could because water offered a natural countermeasure to any electrical/powder-triggered explosive. The canals also provided the platoon a means of cooling, a fact that would become vital after the day's heavy fighting had finished.

The Battle

Immediately following their emergence from the cemetery, the platoon noted a small stream east of and adjacent to their position. The stream ran south and filtered into an east-west canal roughly 25 meters south of their location. A walled grape field sat directly west and adjacent to the platoon. With the knowledge that the location of the airstrike was still a good distance south and west of their position and understanding that the banks of the stream formed a natural path for IED placement, 3d Platoon pivoted west and began making their way through the densely vegetated and walled grape field north of the canal. The westward shift forced the platoon to adjust to a linear formation and pushed Bartlett's and Peck's lead elements between 75 and 100 meters ahead of Hays' squad.

Bartlett's squad and Peck's headquarters element emerged west from the grape field 25 meters north of the canal when the first shots rang out from a point approximately 100 meters south of their position, just north of Route RED STRIPE. Sergeant Hetrick and Private First Class Jacob Brock turned south toward the initial shots when a hail of machine gun fire and RPG blasts filled the air. The Soldiers' training immediately kicked in. They got down and dispersed using individual movement and fire-team bounding. Hetrick and Brock were the first into the canal. Squad automatic

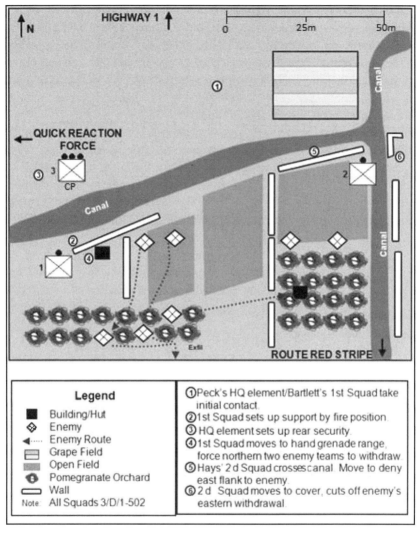

Figure 1. Movement to Contact in Zhari District.

weapon (SAW) gunner Private First Class James Campbell was close behind. Bartlett and the remainder of his squad shifted west and Peck opened fire on everything south. Once Hetrick and Brock hit the canal, Bartlett's squad followed suit.

Hetrick and Brock returned fire as they slogged across the canal, scanning south to locate the enemy machine gun and RPG positions. From 50 meters southeast of their position, hidden in a well-concealed *wadi* situated between an open field and a pomegranate orchard, two enemy teams engaged the Soldiers with mostly inaccurate machine gun and RPG fire. Fire from the west was deadlier and it came from two directions.

The first location comprised two teams of four insurgents armed with light machine guns and RPGs firing from behind a mud wall at the most northern and eastern portion of an open field. From the second location 20 to 30 meters south of the first group, a four-man team of Taliban lay concealed in a pomegranate orchard. Unknown to the platoon, this team possessed one of the most accurate RPG gunners the 3d Platoon had ever encountered. Both Hetrick and Brock made it across the canal to a small mud wall and ducked for cover just as an RPG exploded directly behind them. The blast knocked both Soldiers over. Brock took the brunt of the blast. From his position, Bartlett witnessed the RPG. He later recalled, "We realized that they were pretty frigging accurate with their RPGs…and from what I saw, I thought [Hetrick and Brock] were dead. They survived, they were good."[10] Dazed, Hetrick and Brock steadied themselves, performed a quick check, and continued moving west along the mud wall. As Bartlett and his squad crossed the canal, Hetrick set the men up behind the mud wall's cover. Once set up, Bartlett's squad consisted of himself, Hetrick, Brock, Private Nicolas Haight, Specialist Samuel Poff, and Campbell. The squad let loose a fury of lead toward the insurgents to the south.

Peck witnessed the action south of his location and shifted west, moving parallel to Hetrick's southern elements. As he moved, the platoon leader attempted to radio his commander and describe the engagement that was developing. Staying north of the canal, Peck and his men "were covering distance farther west to make sure no one was coming from [that direction]…At that time, they were unable to raise [Delta] company… due to the terrain."[11] While he was still unable to raise Delta Company, Peck maintained his movement, attempting to overcome the terrain's disabling effect on radio communication. Still in the grape field to the east, Calderaro was also unable to make contact with the Delta Company CP. Situated with Sergeant Hays' squad, Hays' men and Calderaro moved west through the grape field unseen by the enemy.

With possession of the initiative firmly in the insurgents' grasp, the enemy determined the time and place of battle. From the moment the enemy fired its first volley, 3d Platoon was thrust into a fight in which the enemy also had a large amount of control over the tempo and nature of the battle. The Taliban could initiate combat and withdraw from an engagement when it suited them — not necessarily when the battle reached a natural endpoint. This aspect of combat in Afghanistan was a fact of life for coalition forces facing an insurgent forces. Calderaro put it succinctly, "Especially with the enemy we were fighting over there…the enemy is going to do what they think they can win at."[12] When the five Taliban teams opened up with RPG and machine gun fire, they trained their sights on nine US Soldiers and their two interpreters and four ANP personnel. Given the conditions, the odds favored the insurgents. Firmly entrenched behind cover, the enemy maintained a definitive two-to-one numerical advantage, knew the terrain, held the element of surprise, and had preplanned their disengagement routes.

From the Taliban's perspective, every facet of the battle favored them. Every facet save one. The Taliban were unaware that 3d platoon possessed uncommitted forces in the grape field. Because of the field's size and dense vegetation, Hays' squad of six had shifted its maneuver formation, falling 75 meters behind the platoon's lead element. When the enemy opened the battle, the grape field shielded Hays' squad from the enemy. Calderaro stated, "When [the enemy] made the initial contact with Bartlett's lead team, there was a whole other six or seven guys back there that weren't receiving fire."[13] As Bartlett's squad established a support-by-fire position at the west flank, Hays' squad moved uncontested east of the enemy positions. From their vantage point north of the canal, Hays' squad saw the most likely avenues of enemy withdrawal on Route RED STRIPE to the south and east. While Hays' squad began crossing the canal to take position on the eastern flank and with Bartlett's squad already positioned in the west, the platoon shifted its focus toward recapturing the initiative. The men of 3d Platoon were of a single mind: cut off the enemy's retreat, then close with and destroy the insurgents.

In the west, the battle became one of two seasoned boxers jockeying for position. The enemy had opened with a haymaker and nearly connected. Once US Soldiers had crossed the canal and taken cover, the fight briefly slowed to occasional jabs. Describing the situation, Hetrick stated, "we got on the wall and we were just like looking at guys popping up, shooting at us, we'd shoot back at them."[14] With the enemy so firmly entrenched and directing fires at the squad on the west flank, Bartlett's men needed to maneuver to within hand grenade range in order to dislodge the insurgents. Yet, in addition to the intense incoming fire, the squad faced terrain that was potentially mined. Hetrick stated, "We're just moving along the canal you know, pretty slow because no one had been down there. We weren't sure what was around the next corner. We're also thinking about IEDs because it'd be a perfect place for dismounted IEDs along that canal."[15] Their advance was understandably methodical.

The squad's attempt to maneuver served as a catalyst. Aware that Bartlett's men were advancing, the enemy responded like a stirred hornet's nest, unleashing a torrent of machine gun fire and RPG explosions to pin the Americans in place. The Taliban were fighting Bartlett's squad for every inch of terrain. Hetrick recognized that the enemy force possessed fire superiority and outnumbered his element. He said, "They had about eight guys and very accurate RPG fire. We started to take multiple RPGs. I'd say like one coming in every 15 to 30 seconds. It seemed like every time you would move up, they would respond with an RPG."[16] Against the oppressive rain of fire, Bartlett and his men threw smoke grenades to advance. When those ran out, they threw fragmentary grenades.[17] Beyond the fragmentary grenades' masking potential, the successive explosions forced the enemy to keep their heads low, decreasing their rate of machine-gun and RPG fire. Additionally, the grenades pinned the enemy in place, preventing them from displacing to take up better firing positions. As Bartlett's squad inched along the western edge of the wall, they pivoted around the wall's edge and raced to a nearby small hut on the wall's south side. Still attempting to push the enemy back, a few of the Soldiers threw grenades which arced perfectly toward their targets. The northern Taliban teams were forced to withdraw south, racing across an open field to link up with the team in the pomegranate orchard.

The time was 0945. It was just 15 minutes into the battle and the Taliban were down two teams. With their retreat, the enemy lost a significant amount of suppressive firepower including small arms, RPGs, and light machine guns. The character of the fight now decidedly shifted in favor of US forces. Hetrick recognized this fact later, stating, "I do not believe that we really gained the upper hand in the firefight until myself and Sergeant Bartlett's squad had fully pushed south of the canal and maneuvered to within hand grenade range of the Taliban."[18] For the enemy, their situation was grimmer than even the US forces realized. The northern teams withdrew not only because Bartlett's squad had compromised their position but also because they had suffered significant casualties. Their morale broken, the Taliban teams were actively looking to disengage even as their sister team fought on. Little did the enemy know the battle's dynamic was about to shift in even greater favor of the US platoon. Calderaro had made radio contact with Delta Company. He was told that the 4th Platoon's assets were on their way to the battlefield and CCA would be on station soon thereafter.

With the northern enemy teams retreating before Hetrick and Bartlett, the fury of battle briefly dimmed. The first phase of battle came to a close as both sides collectively caught their breath. Hetrick and Bartlett's squad moved and cleared the small structure. Calderaro crossed the canal just north of the enemy's central position. Hays' squad crossed as well, coming to a halt adjacent to Calderaro. From behind the cover of a small mud wall north of an open field, they examined the eastern portion of the battlefield for a route of maneuver to cut off the enemy's eastern withdrawal. On the other side of the canal north of Bartlett and Hetrick's position, Peck was in communication with the QRF vehicles.

Third platoon did not have to wait long. At 1000 the vehicle element of the QRF force arrived directly west of Peck's position. Carrying 4th Platoon, the QRF consisted of four mine-resistant armored vehicles. A fearsome sight, the vehicles brought a total of 18 infantryman and were armed with one MK-19 grenade launcher, two M-2 .50-caliber machine guns, and one M240 machine gun, respectively.[19]

The QRF's show of force broke the Taliban's will to resist on the western side of the battlefield. Taking advantage of the temporary lull in combat, the RPG team dashed east as fast as possible, moving to the central pomegranate orchard south of Calderaro's position to link up with the remaining enemy teams and take cover in a well-concealed, lean-to shelter in the orchard. Injured, the northern teams disengaged and withdrew south to Route RED STRIPE, moving west toward the town of Kandalay. Though their heart was no longer in the fight, Hetrick noted that the men still disengaged in a manner advantageous to them, "From my fighting position at the time, I could see them moving from cover to cover south and then once they reached Route RED STRIPE, they almost all turned and fired a long burst, perhaps the rest of their ammo. They [withdrew] in good order, always covering their moves with fire, not simply just running away."[20]

Emptying their clips and tossing or stashing their weapons in hidden caches, within minutes, the Taliban were able to transform from combatant to non-combatant. To complete the transformation, the enemy kept caches of farming clothes nearby to quickly change from insurgent to civilian.[21] Additionally, the Taliban stored wheelbarrows and motorcycles nearby for the wounded and weapons, respectively. Ultimately, the Taliban understood that if they could create situations where doubt persisted in the minds of US Soldiers, the latter would honor the rules of engagement and grudgingly permit the Taliban fighter his escape.

Hetrick and Bartlett owned the western flank and with the southern RPG team moving east into the center position, they sensed that Calderaro and Hays' positions were about to come under fire. Peck was communicating with the QRF force but would soon be moving his element to link up with Bartlett for a joint push south to cut off any further retreat west along Route RED STRIPE. This gave Hetrick the opportunity he had been waiting for, to provide the hammer to Hays' anvil. Hetrick would shift east, link with Calderaro and, secure in the knowledge that the enemy's eastern escape route was cut off by Hays' squad, he and his grenadiers would provide yellow smoke for the QRF to pound the pomegranate orchard with enough explosives to litter the orchard with corpses. Should that fail, Hetrick was prepared to supply the necessary firepower himself. To achieve this objective and in preparation for linking up with Calderaro, Hetrick gathered grenadiers Brock and Haight and the SAW gunner Campbell, leaving Bartlett with rifleman Poff.

As Hetrick began his eastward movement to Calderaro's position, Bartlett patiently waited for Peck's squad to join with him and Poff so they could begin their move south to Route RED STRIPE in the west. With Hetrick acting as the hammer and Bartlett closing the western avenue of retreat to the enemy, the only way for the enemy to escape from the pomegranate orchard south of Calderaro was to the east, covered by Hays' squad. Possessing an innate talent to read the battlefield and having traversed to his position largely unnoticed by enemy forces, Hays began to shift his squad east of Calderaro to take control of the eastern flank. In addition to himself, Hays's squad consisted of his machine gunner Specialist Brook McQuay, riflemen Private Brian Dickson and Private First Class Ethan Bowe, and grenadier Private Delton Graves. Setting his men up on line facing south, Hays prepared to close the enemy's sole remaining avenue of retreat. Always generous in his praise of his NCOs and their men, Peck stated, "Having the aggressive NCOs that I had was a huge asset and I try very hard never to reign them in too much."[22]

Peck's willingness to trust in his NCOs to develop the situation paid dividends in the platoon's maneuver. While Peck communicated with the QRF vehicles, Calderaro stated that Hays saw "on the south side of the creek...three big open fields that...were flat fields. So [Hays] pushed to the far east of it and then [Calderaro] set the machine gun up in the middle."[23] From his vantage point, Hays possessed a clear line of sight across open fields south to Route RED STRIPE, the road that the western Taliban elements had made their withdrawal on moments earlier. Immediately to the east of the orchard ran a north-south creek with walled grape fields immediately across it. With those walls serving as effective obstacles and with Hays' concentration of firepower aimed south over the open fields and creek, Hays' position created a no-man's land guaranteeing that any

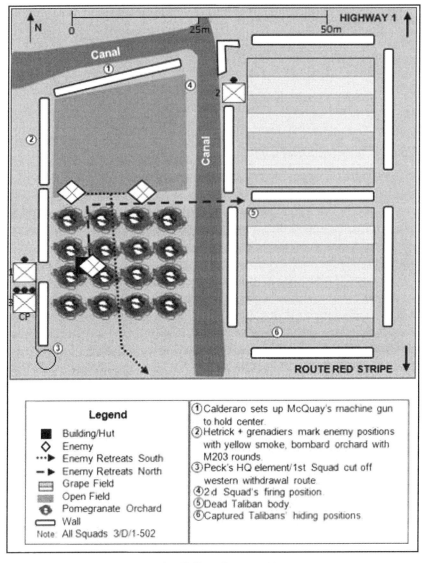

Figure 2. The Taliban Attempt Disengage.

east-ward retreating enemy would run straight into a killing zone. As it turned out, this is precisely what happened.

Ready to drop the hammer, Hetrick, Brock, Haight, and Campbell maneuvered east toward Calderaro. Sensing the noose was tightening around the pomegranate orchard, the three remaining Taliban teams unleashed a torrent of machine gun and RPG fire toward Calderaro, Hays, and Hetrick's positions. The entire 3d Platoon responded in unison. While

Hays' squad returned fire in the east from Peck's position northwest, Tate and "Doc" Rankin poured fire into the orchard. At Calderaro's position, Hays' machine gunner McQuay sent a wall of suppressive fire down the center. Meanwhile, Hetrick, Brock, and Haight responded with an onslaught of their own, firing successive volleys of grenades from their M-203s into the orchard. By the end of the firefight, the grenadiers had fired roughly 80 40-mm rounds, most from this position.[24] Against the barrage of ordnance, enemy fire fell significantly.

The Taliban scurried for cover along intersecting *wadi* lines, searching the area for safe haven against the rush of incoming fire. When Calderaro saw the enemy take up positions behind 12-foot mud walls, he called Hetrick to adjust the grenadiers' fire to target the mud walls themselves.[25] The grenadiers continually tried to keep the enemy off balance, homing in on fixed enemy positions and engaging them in order to kill them, break their cover, or force them to adjust their positions. Hetrick and Peck both assessed the grenadiers' performance as vital to the fight with Hetrick stating that the M-203 was the most pivotal weapon the platoon had.[26] For his part, Peck argued that the grenades caused the enemy to keep their heads down and prevented them from moving against the platoon.[27] Once Hetrick made the face-to-face linkup with Calderaro, Calderaro ordered Hetrick to have his grenadiers fire yellow smoke into the pomegranate orchard in order to mark locations for the QRF's vehicle mounted weapons. Hetrick dashed back to his men, throwing several hand grenades while on the run. There he instructed his men to fire yellow smoke from the 203s.

Unfortunately for the 3d Platoon, from the moment the QRF vehicles set up north of the canal, radio malfunctions and poor positioning caused significant difficulties. The problems began when someone within the 4th Platoon's QRF force held down the talk button on their handset creating a "hot mic" situation, a condition in which all message traffic is blocked by the open handset. When it became clear that he would be unable to communicate with the vehicles over the radio, Peck sprinted through what he later referred to as "ineffective machine-gun fire" in order to relay the message calling for the fire he wanted.[28] Hetrick and his grenadiers had already marked the enemy positions in the central pomegranate orchard with yellow smoke but the machine gun and the MK-19 gunners on the vehicle could not see the smoke. Peck made a second run for the QRF vehicles to see if there was any way he could adjust the vehicles or their fires to support his platoon, but it was to no avail. The vehicles would be of no further use in the battle.

At 1015, two key events occurred while Peck was on his second run back to the QRF vehicles. The first was notification that two OH-58 Kiowas and one AH-64 Apache helicopter were on station. Possessing superior sensors to the Kiowas, the Apache floated above the fray, ready to serve solely as the 3d Platoon's eyes in the sky while the Kiowas, which were equipped with .50-caliber machine guns and 2.75mm rockets, provided direct fire support to the platoon. Recognizing that the QRF vehicles were going to be of no use to the platoon from their current position, Peck asked the Kiowas to switch to the platoon net so that he could communicate directly with the pilots. He was about to return to his headquarters element when he ran into Sergeant James Brafford, the company RTO. Isolated from the QRF by the series of microphone miscommunications, Brafford asked Peck if he could use some help. Peck grouped Brafford with his own element and immediately set off to find Bartlett and Poff. Brafford then took point as Bartlett began moving south to Route RED STRIPE to cut off the enemy's western route of withdrawal.[29] Peck later offered Brafford accolades, stating that he was a "great adaptive Soldier that integrated in with my guys and ran himself into the ground with the rest of us to capture the bad guys."[30] True to Peck's words, Brafford remained with the 3d Platoon for the entire action.[31]

Within the lean-to, the Taliban RPG team now considered the battle lost. There can be little doubt that the team's attempt to escape was largely influenced by the sight of Peck's headquarters element crossing the canal northwest, the enormous rush of firepower from Hetrick, Calderaro, and Hays' positions, and the whirring of US helicopters overhead. The RPG team used the *wadi* lines north of the orchard for as long as possible before emerging east from the pomegranate orchard. Once in the open, they made a dash for the creek bed and to the walled grape field beyond when they were engaged by McQuay's machine gun and Hays' squad. From Hays' position, Bowe zeroed in, took aim, and pulled the trigger. The bullet entered the RPG gunner's cheek and exited the back of his head, taking a significant portion of his brain with it. He died instantly.[32] Though the US Soldiers did not know this at the time, Bowe had just killed the local enemy leader. The three remaining team members dragged him into the adjacent grape field. Still under fire, the insurgents abandoned their efforts, scattered the team's weapons, and hid underneath a bunch of grape vines at the southernmost point of the grape field closest to Route RED STRIPE. Demoralized and disorganized by the loss of their leader, they intended to escape east along Route RED STRIPE as soon as the right moment presented itself.

That moment never came. As soon as the Kiowas came on station, Calderaro sent out a message platoon-wide for everyone to pull out their VS17 panels to mark their positions.[33] The men of 3d Platoon reacted by placing the six-foot bright red and orange panels in front of their positions. Speaking with Calderaro on the platoon net, the Kiowas began their attack runs. Vectoring in .50-caliber and 2.75mm rockets on the lean-to and the northern *wadi* lines, the Kiowas pummeled the pomegranate orchard with a storm of steel and explosives.

For both the Taliban hiding in the grape field and those being battered in the orchard, retreat was now exceedingly difficult. Calderaro knew Route RED STRIPE was the Taliban means of escape. In addition to the Kiowas' gun runs he had the helicopters patrolling the Taliban routes of withdrawal, boxing them into the orchard. Nevertheless with casualties mounting, the two remaining teams of Taliban had to chance escape or face extinction. With wheelbarrows, motorcycles, and in one insurgent's case, a comrade's shoulders at the ready, the two remaining teams prepared to beat a hasty retreat.

From their vantage point, the helicopters witnessed the withdrawal but were unable to engage as the insurgents possessed no visible weapons. Before the remaining Taliban withdrew south and east down Route RED STRIPE, they placed their weapons in a covered wheelbarrow and laid a sprawled comrade in the other. Given the blood markings on the man's shirt and the way that he was positioned in the wheelbarrow, the helicopters concluded the man was dead. Still another Taliban held tightly onto an insurgent's back as he carried him off the battlefield. Whether because he believed that he had placed enough distance between himself and US forces or because he simply could not bear the man's weight any longer, the insurgent placed the unconscious man underneath a tree on the side of the road.

The helicopters relayed to 3d Platoon that this insurgent showed no movement. As with the man in the wheelbarrow, he was determined to be dead.[34] Peck requested permission from Delta Company to give pursuit and retrieve the possible dead insurgent but was denied. Even given the short distance, the threat of IEDs along Route RED STRIPE was too great. The 3d Platoon had won the day without a single casualty and had inflicted what the battalion S2 later estimated were three enemy killed in action and seven wounded in the process. That would have to do for the day's fighting. Following the action, a rash of funerals in the area suggested that the initial tally for enemy casualties may have erred on the conservative side.

After the Battle

With fighting over by 1100, the platoon began the process of site exploitation on the pomegranate orchard and picking up the dead Taliban. In so doing, the Soldiers also accomplished a rare and dramatic feat in the action's waning hours, the capture of three insurgents. Hetrick, Calderaro, Bowe, and Campbell approached the orchard from the west while Hays, Dickson, and Graves approached from the east. Still wary of potential hazards, US forces moved to the orchard using fire teams in alternating bounds. Upon arriving at the lean-to, Hetrick placed Campbell in the northwest corner on overwatch south while Hetrick and Bowe entered from the east. The Taliban had placed dried grass and grapevines on top of the lean-to, giving them excellent concealment from aerial surveillance.

Further south, 3d Platoon discovered a large bed-down area complete with partially eaten meals and unpoured hot chai tea. The evidence was clear that the Taliban were surprised by the 3d Platoon's initial approach. Even more surprising was the size of the bed-down site. It was spacious enough for more than 30 insurgents, contained perfect fields of fire, and was littered with components used in the creation of HME and booby traps to surround the field at night while the men slept. This location likely provided crucial support to the illegal Taliban checkpoint and most likely served as an insurgent command and control node. With the pomegranate orchard, the bed-down site, and the bomb making caches in US hands, 3d Platoon moved to pick up the insurgent Bowe had slain.[35]

At 1120, 3d Platoon picked up the dead Taliban's body. The evidence found on him indicated that he was a local Taliban commander. In addition to the Pakistani cash and ICOM hand-held radio found on his corpse, he was also older than his fellow insurgents, wore a load-bearing vest, and fought with greater RPG accuracy than the platoon had ever seen. For propaganda purposes, he also had scraps of paper with Koranic verses sewn into his clothing. Third Platoon strongly believed that the insurgent had been trained in Pakistan. In the platoon's eyes, this was likely a hardened Taliban fighter sent to command local insurgent forces.

The helicopters had tracked the three insurgents hiding in the grape field. Though the Kiowas had lost visual contact, they gave Peck's platoon an idea where to look. Bartlett, Hetrick, Poff, Brafford, and Graves were moving through the grape fields, each man taking a single row, when Bartlett and Hetrick made brief visual contact with an insurgent and gave chase.[36] Though running on adrenaline, the Soldiers also ran with more than 100 pounds of equipment on their bodies. The temperature was an unrelenting 135 degrees and to this, the grape field added excessive humidity. Inasmuch as the day's combat was an exercise in physical exhaustion, Hays later remarked that "the biggest smoker was actually running through the fields."[37] Despite the severe climate and terrain conditions, a far greater danger lurked within the sinuous foliage. Dug into deep eight-foot trenches, the plant grew at the base of huge dirt rows, producing a dense vegetation wall not unlike the hedgerows US Soldiers found in northwestern France during World War II. As in Europe in 1944, the vines in the grape fields easily concealed fighting positions and explosive devices and, in this case, provided the enemy a safe haven to hide for possible retreat. As the fleeing insurgents bobbed and weaved to elude US troops, Bartlett, Hetrick, and Poff lost visual contact for what was no more than five to ten seconds. This was all the time the insurgents required to disappear. Watchful that they might be about to walk into their second ambush of the day, the men spread out and began a slow and deliberate search.

The grape rows limited the pursuing US troops to three and a half feet wide corridors with spacing between the vines even narrower still. The area was so constrictive it prevented US Soldiers from training their weapons on the enemy. Hetrick believed the Taliban had successfully withdrawn when he spotted two AK-47s leaning against the grape wall and promptly called out, "We got weapons up here!"[38] As he continued walking forward, Hetrick spotted underneath the foliage the whites of two pairs of Taliban eyes staring in petrified shock back at him. Falling back on training and instinct, he pulled the one insurgent out, subdued him, and pulled the other out for Bartlett and Poff to detain. As the troops moved to the next grape row, they spotted and subdued the final insurgent who was hiding with his PKM.

In addition to the three men captured, 3d Platoon also captured two AK-47s, one pistol, one RPG with rocket, one PKM with an empty 200 round drum, two magazines, and two chest racks. They had in their possession numerous IED manufacturing and intelligence materials to include pressure plate materials, four sim cards, a motorcycle title and receipt, a prescription note from a Pakistani hospital, and two Arabic Koran verse sheets among other items.[39]

After detaining the insurgents, the ANP, according to Hetrick, "seemed to show up out of nowhere, where they began speaking with the [enemy prisoners of war]."[40] In spite of overwhelming evidence to the contrary, the ANP, after speaking for a brief time with the captured insurgents, contended that the detainees were not insurgents but local farmers. Peck's interpreters were put to good use questioning the three detainees. When the interpreters relayed to Bartlett, Hetrick, Brafford, Poff, and Graves that the ANP claimed the captured enemy were armed farmers, all five men had a good laugh. The platoon ignored the ANP and had them assist in moving the detainees to a more secure location.

Figure 3. Captured Enemy with their Equipment.

Lieutenant Peck placed the captured Taliban along a bend in the creek and assigned his men to positions where they could provide security. From their position, 4th Platoon briefly maneuvered their vehicles west before heading south toward Route RED STRIPE to provide additional overwatch. The vehicles moved methodically because 4th Platoon was warned that Route RED STRIPE was a heavily mined road. By this point the radio network was working and again and Peck called his commander, requesting assistance with the detainees. He received word that 1st Platoon was dispatched to secure the detainees and bring them to FOB Wilson for processing. For Peck and the 3d Platoon, the greatest threat now lay in the fact that the platoon was "black" on water, meaning the Platoon's water supply was now below 25 percent. Peck asked that 1st Platoon bring with it a resupply of water rations immediately. Emphasizing the severity of the situation, Peck stated, "I mean no one had any water."[41]

Technically, Peck's platoon still was committed to conducting the BDA mission. Yet while Peck, Bartlett, Calderaro, and Hetrick performed site exploitation, picked up the dead Taliban, and captured the enemy,

Hays watched as the adrenaline wore off the remainder of the platoon and fatigue set in. Speaking with Peck, both men agreed that the combination of continued operations and the elevated temperature took a serious toll on the men in the platoon.

Peck then ordered that the small amount of water remaining be redistributed. He also issued power bars and granola to try to put salt back in the men's system. Additionally, he had men set up security in the shade and rest neck deep in the canals, taking off as much of their gear as possible. However, after the weight of a seven hour patrol and a two and a half hour firefight, fatigue carried the day. Peck assessed the platoon combat ineffective.[42] According to Bartlett, Peck "was requesting people to come out and replace our Soldiers because at least 95 percent of the platoon was throwing up, [had] heat cramps, and just couldn't go no more. Everybody was going down."[43] Peck had four severe heat casualties requiring an IV or a rehydration salt packet.[44] Always glib, Calderaro stated that "if [the Taliban] decided to counterattack us at this time, we would have been some screwed frigging puppies."[45] All told, even though Peck's platoon had stayed within a two kilometer radius, they estimated that they had traversed somewhere between 8-10 kilometers that day.

War is never short on irony. Before the Kiowa helicopters left to refuel, the pilots radioed Peck and asked if he and his platoon needed any more assistance. Peck replied that if they had any water, he could use it. This exchange led to a lighter moment fitting for the 3d Platoon, in that it still involved an element of grave danger. The Kiowas flying some 60 feet overhead began runs to drop heavy bundled packages of water bottles to the ground at excessive speeds. Physics took hold of the packages and the bundles hurtled to the earth with tremendous force. Few water bottles survived. Sergeant Nicholas Hays was standing underneath a tree when one of the packages plummeted toward him. Luckily for Hays, the package smashed near the tree top. The resulting explosion provided Hays with a deathly scare and a welcome shower. One of the water bottles survived the ordeal and Hays picked it up. According to Brafford, "Hays got a bottle of water out of the deal. He came back with his bottle of water, his eyes all big and white and hands shaking, he was like, 'I got water, but I almost died.'"[46]

As 1st Platoon arrived to take away the detainees, 4th Platoon dismounted their vehicles to take over the task of performing the battle damage assessment for 3d Platoon, who would now take 4th Platoon's vehicles and head back to COP Senjaray once their site exploitation was complete. Giving the platoon over temporarily to Sergeant First Class

Calderaro, Peck traveled with 4th Platoon to conduct the BDA. While on foot to the airstrike location, 4th Platoon received sporadic harassing fire but found no major resistance. When they finally reached the location, 4th Platoon found the HME factory destroyed. Carnage around the site indicated six to eight individuals once worked there. There was an extremely strong chemical smell that caused some members of the platoon to vomit and their eyes to burn. They pulled back and took pictures from a safe distance.[47] When Peck finally reunited with his men for the ride back to COP Senjaray, his driver's legs were so badly cramped that Peck had to use the barrel of his M4 to hit the gas and brakes. In effect, the dull thudding sound of Peck's M4 against the vehicle's pedals served as a final exclamation point to the day's action, a powerful indicator of the 3d Platoon's exhaustion.[48]

Aftermath

After the firefight, 3d Platoon's members stated that they were "never tested again at the same level" as they were on 23 June.[49] Their tenacity in the face of heavy firepower, their willingness to maneuver, to close with, to pursue, and even to hunt down the Taliban quietly garnered a reputation with the enemy. During the 1-502 IN's last two and a half months in Afghanistan, 3d Platoon, Delta Company received intelligence reports that the Taliban were targeting a group of individuals led by a man with a big knife on his body armor. Peck was the only Platoon Leader in the 1-502 IN to carry a machete on his body armor. He took the news in stride proudly stating, "I had a bounty on my head."[50]

Yet for all its boldness on the battlefield, the platoon remained versatile in the face of the dynamic complexities that Afghanistan presented. Competent leadership from perceptive and adaptive NCOs expanded the platoon's capability tremendously and formed the foundation underneath the success that day in Zhari. Their engagement in daily "white pages" missions led them to embrace the adaptability required for a COIN effort with Peck noting the need to be ready to take the fight to the enemy even within an environment dominated by non-lethal operations.

When engaged in lethal operations, US units are accustomed to having a variety of munitions and fires at their disposal. Yet on 23 June, 3d Platoon operated without mortar or artillery support and without the immediate fires that the QRF was supposed to bring to bear. The presence of the OH-58 Kiowas was a welcome addition but they did not arrive until 3d Platoon had already shifted the battle in favor of US forces. Victory on that day was a result of the platoon's leadership's ability to understand the battlefield and use maneuver decisively. The platoon's aggressive and perceptive leaders made contact with the enemy, developed the situation, and gained the initiative, allowing for an almost complete defeat of the insurgent force they faced. Also critical to this success was the platoon's collective conditioning, mental toughness, and endurance which proved equally as important as any weapon system. In summation, decisiveness and adaptability are two of the qualities most needed for the US Soldier to be successful on the modern battlefield.

Third Platoon embodied these sentiments. For them, it was the equivalent of flipping a switch, to be able to shift immediately from security operations to the attack. As Sergeant Bartlett stated, "we'd have to flip that switch. You know, one second you're just calm, you're just pulling security, you're not showing you're a hard target and you just flip that switch, and it's frigging on, you're ready to go."[51] On 23 June 2010, 3d Platoon flipped that switch and every Soldier survived unharmed. It was a process they would repeat for the remainder of their stay in Afghanistan and to their credit this method produced the same result each and every time. During their time in Afghanistan, 3d Platoon suffered no combat related injuries. Everyone from the unit returned home safely.

Notes

1. Sergeant Michael Calderaro, quoted in 3d Platoon, Delta Company, 1-502 IN IN group interview by Michael Doidge, Combat Studies Institute, Fort Leavenworth, KS, 28 June 2011, 25.

2. Lieutenant Evan Peck, quoted in 3d Platoon, Delta Company, 1-502 IN IN group interview by Michael Doidge, Combat Studies Institute, Fort Leavenworth KS, 28 June 2011, 2.

3. Peck, quoted in 3/D/1-502 IN group interview, 44.

4. Sergeant Matthew Hetrick, email to Michael Doidge, Combat Studies Institute, Fort Leavenworth, KS, 28 July 2011.

5. Hetrick, email, 18 July 2011.

6. Sergeant Matthew Hetrick, quoted in 3d Platoon, Delta Company, 1-502 IN IN group interview by Michael Doidge, Combat Studies Institute, Fort Leavenworth, KS, 28 June 2011, 52.

7. Calderaro, quoted in 3/D/1-502 IN group interview, 7-8.

8. Lieutenant Evan Peck email to Michael Doidge, Combat Studies Institute, Fort Leavenworth, KS, 23 July 2011.

9. Hetrick, email, 18 July 2011.

10. Sergeant Daniel Bartlett, quoted in 3d Platoon, Delta Company, 1-502 IN IN group interview by Michael Doidge, Combat Studies Institute, Fort Leavenworth, KS, 28 June 2011, 9.

11. Peck, quoted in 3/D/1-502 IN group interview, 14.

12. Calderaro, quoted in 3/D/1-502 IN group interview, 57.

13. Calderaro, quoted in 3/D/1-502 IN group interview, 10.

14. Hetrick, quoted in 3/D/1-502 IN group interview, 13.

15. Hetrick, quoted in 3/D/1-502 IN group interview, 13.

16. Hetrick, quoted in 3/D/1-502 IN group interview, 13.

17. Peck, quoted in 3/D/1-502 IN group interview, 14.

18. Hetrick, email, 18 July 2011.

19. Peck, email, 23 July 2011.

20. Hetrick, email, 18 July 2011.

21. Sergeant James Brafford, interview by Michael Doidge, Combat Studies Institute, Fort Leavenworth, KS, 4 August 2011, 18.

22. Peck, quoted in 3/D/1-502 IN group interview, 56.

23. Calderaro, quoted in 3/D/1-502 IN group interview, 15.

24. Hetrick, email, 18 July 2011.

25. Hetrick, email, 18 July 2011.

26. Hetrick, email, 18 July 2011.

27. Peck, email, 23 July 2011.

28. Peck, email, 23 July 2011.

29. Brafford, interview, 4.

30. Peck, email, 23 July 2011.

31. Peck, email, 23 July 2011. Brafford, interview, 32.

32. Brafford, interview, 10.

33. Calderaro, quoted in 3/D/1-502 IN group interview, 16.

34. Hetrick, email, 18 July 2011.

35. Peck, quoted in 3/D/1-502 IN group interview, 28.

36. Calderaro, quoted in 3/D/1-502 IN group interview, 26. Hetrick, quoted in 3/D/1-502 IN group interview, 32. Bartlett, quoted in 3/D/1-502 IN group interview, 32.

37. Sergeant Nicholas Hays, quoted in 3d Platoon, Delta Company, 1-502 IN IN by Michael Doidge, Combat Studies Institute, Fort Leavenworth, KS, 28 June 2011, 54.

38. Hetrick, quoted in 3/D/1-502 IN group interview, 32.

39. 2d Brigade Combat Team, 101st Airborne Division, "BDA/SSE Assessment of Airstrike, CTF Talon D/1-502, 23 June 2010," *After Action Report*, 23 June 2010, slide 5.

40. Hetrick, email, 18 July 2011.

41. Peck, quoted in 3/D/1-502 IN group interview, 37.

42. Peck, quoted in 3/D/1-502 IN group interview, 40.

43. Bartlett, quoted in 3/D/1-502 IN group interview, 40.

44. Peck, quoted in 3/D/1-502 IN group interview, 54.

45. Calderaro, quoted in 3/D/1-502 IN group interview, 55.

46. Brafford, quoted in 3/D/1-502 IN group interview, 31.

47. 2BCT, 101st Airborne Division "BDA/SSE Assessment of Airstrike," 23 June 2010, slide 1, 3.

48. Peck, quoted in 3/D/1-502 IN group interview, 54.

49. Peck, quoted in 3/D/1-502 IN group interview, 56.

50. Peck, quoted in 3/D/1-502 IN group interview, 60.

51. Bartlett, quoted in 3/D/1-502 IN group interview, 52.

Forging Alliances at Yargul Village
A Lieutenant's Struggle to Improve Security
by
Anthony E. Carlson, Ph. D.

In March 2010, Sher Khan, the tribal elder of Yargul village, stormed onto Forward Operating Base (FOB) Wright on the outskirts of Asadabad, Afghanistan. Upon arriving at FOB Wright, Sher complained to the Kunar Provincial Reconstruction Team (PRT), a military-civil unit dedicated to improving Afghan governance and stability, that recurrent flooding damaged Yargul homes, crop fields, and roads. The source of the flooding was a small aqueduct. Originating in the lower Pech River valley, the aqueduct delivered water to Asadabad, Yargul, FOB Wright, and nearby irrigated farms. As part of an effort to reduce improvised explosive device (IED) casualties, the US military had installed steel grates over canal, road, and aqueduct culverts throughout Kunar.[1] "If someone stuck an IED in [the canal], where the kids play," observed Sergeant First Class Christina LeMond, the PRT Civil Affairs Team's (CAT) purchasing officer, "a lot of people would be hurt."[2] The culvert denial grates disrupted the concealment of IEDs. Yet because Afghans discarded trash and other debris into waterways, garbage flowed into the grates during heavy rainstorms. At Yargul, which was located literally next to FOB Wright, trash flowed into the grate during downpours, creating a makeshift dam and causing a crest of water that submerged the village.

During his meeting with the Kunar PRT Chief US Navy Commander Mark Edwards, Sher argued that it was the PRT's responsibility to oversee the removal and disposal of trash from the aqueduct. "We're flooded. You didn't clean the grate...Now we've lost farmland. Our houses are filled with water," Sher bristled. "It's your responsibility to come down and clean this thing because it's part of the base."[3] Edwards sympathized with the plight of Yargul farmers.

Given the proximity of Yargul to FOB Wright, it behooved coalition forces to cultivate goodwill with Sher avoid alienating his villagers, and deter them from supporting the Taliban insurgency. As a result, Edwards directed the Kunar PRT CAT to study the problem and recommend a solution.

Kunar PRT CAT Project Manager First Lieutenant Joseph Ladisic assumed responsibly for solving the flooding problem. Acting with initiative and a sense of urgency, Ladisic evaluated a diverse range of solutions before implementing a cash-for-work program funded by the Commander's Emergency Response Program (CERP). Administered by

Figure 1. The culvert denial grate outside of FOB Wright (note the trash accumulation on top of the grate).

Figure 2. The aqueduct near Yargul.

US Forces-Afghanistan (USFOR-A), CERP was a Department of Defense program that distributed money to US military commanders for urgent reconstruction, relief, and humanitarian projects in support of overseas stability operations.[4] Utilizing CERP funds, Ladisic hired Sher to pay villagers to remove the trash, dispose of it properly, and keep the grate unobstructed. By authorizing the program, Ladisic intended to eliminate flooding, provide an infusion of cash into Yargul, elevate Sher's status as a local powerbroker, and prevent security incidences at FOB Wright. Endorsing the cash-for-work program, Major Brian Elliott, the PRT CAT's civil-military operations center (CMOC) officer in charge (OIC), explained that "the effect that we were trying to achieve had nothing to do with the clean grate. Ultimately, what we wanted was to make sure that the people that lived next to the wall weren't going to throw a grenade over... we really really don't want to upset these guys that live right next to our wall."[5]

The three month program, which lasted from April to July 2010, yielded mixed results. Although the program reinforced Sher's prestige and power, it also fostered tension, jealousy, and acrimony among rival villages along the aqueduct that were not invited to participate. It also failed to persuade Yargul residents to assume responsibility for flood control. Once the program concluded in July, trash began to accumulate over the grate again. Nevertheless, the cleanup program contributed to the stable security situation around FOB Wright in the first half of 2010. The stability resulted from Ladisic's purposeful leadership and adeptness at forming partnerships with Afghan elders to institute small development programs that, however well intentioned, encountered unanticipated consequences and uneven results.

Background

Kunar Province is located in northeastern Afghanistan alongside the mountainous and porous Pakistan border. Encompassing 1,900 square miles, Kunar is approximately the size of Rhode Island. Given the harsh terrain and absence of integrated roads, Kunar's narrow river valleys serve as arteries of commerce, transportation, and settlement. Rising in the Hindu Kush mountain range in northwestern Pakistan, the Kunar River and its tributaries constitute the major river system of the province. The river enters Afghanistan in Kunar's Nari District before turning southwest, flowing parallel to the Pakistan border, and merging with the Kabul River east of Jalalabad. With a population of 11,000 residents, the provincial capital of Asadabad lays at the confluence of the Kunar and Pech Rivers in east central Kunar Province. The remaining 390,000 Kunar residents, approximately 96 percent of the population, are scattered throughout the rural river valleys, with the main concentrations distributed south of Asadabad. Agriculture is the primary source of revenue for three quarters of Kunar families.[6]

Kunar's rugged and towering terrain, isolated valleys, location as an insurgent smuggling route into Afghanistan, and inward looking tribes contributed to a strong and resilient insurgency that found a sanctuary in the province. The valley of the Korengal River for instance, has been one of the bloodiest places in Afghanistan. Daily clashes between US Soldiers and Taliban insurgents in the six-mile long valley, known as the "Valley of Death," accounted for 20 percent of all combat incidents in Afghanistan before the United States abandoned the valley in April 2010.[7] Ten months later, US forces temporarily withdrew from the Pech River valley, a strategic east-west trade corridor, because of uneven progress, a hostile local populace, and the deaths of 103 US Soldiers since the initial invasion of Afghanistan in 2001. The poor provincial security situation and the population's indifference towards the Government of the Islamic Republic of Afghanistan (GIRoA), throughout 2010, defied simple solutions.[8]

The Kunar PRT was responsible for expanding and deepening the legitimacy of GIRoA in the restive province. The PRT concept emerged in late 2002 as part of the coalition campaign transitioning from military operations against the Taliban and al-Qaeda to a modified effort that focused on building a new Afghanistan. Emphasizing political stabilization, economic development, reconstruction, and the legitimization of Afghan governing institutions, the new coalition campaign introduced PRTs.[9] The teams traced their origins to Coalition Humanitarian Liason Cells, also known as "chiclets". These were groups of Special Forces and military civil affairs personnel. In early 2002, chiclets functioned in major provincial towns and areas of strategic significance. They launched small development projects, assessed humanitarian needs, forged alliances with local powerbrokers, and cooperated with nongovernmental organizations (NGOs) and the United Nations Assistance Mission in Afghanistan (UNAMA).[10] PRTs borrowed from the chiclet model and assumed many of its objectives. Comprised of military and civilian personnel such as representatives of the United States Agency for International Development (USAID), State Department officials, and Department of Defense civilian employees, PRTs sought to improve security, expand GIRoA's authority, and promote economic opportunities. In November 2002, the first PRT was established in Gardez, the provincial capital of Paktia. By the summer of 2004, a dozen PRTs operated throughout Afghanistan with the highest concentrations occurring in southern Afghanistan.[11] In October 2006, the

International Security Assistance Force or ISAF, which is the NATO-led coalition authority, took control of all PRT operations.[12]

In order to respond to diverse local conditions, PRTs enjoyed flexibility, adaptability, and autonomy. Using CERP and US Overseas Humanitarian, Disaster, and Civic Aid (OHDACA) funds, PRTs launched development, infrastructure, water, electrical, sanitation, agriculture, animal husbandry, and governance projects tailored to achieve specific local outcomes. By 2011, 27 PRTs operated throughout Afghanistan.[13]

Genesis of the Cash for Work Cleanup Program

Sher Khan's visit to FOB Wright, in early 2010, presented an opportunity for the incoming Kunar PRT CAT to promote stability and security in the southern hinterland of Asadabad. Composed of US Navy Sailors, US Army Reservists, National Guard Soldiers, and four US government civilians, the Kunar PRT was an ad hoc organization. The CAT, which consisted of seven US Army Reservists and one National Guard Soldier (First Lieutenant Ladisic), served as the PRT's "clearinghouse" for project development and exercised oversight responsibilities for governance issues. Under the leadership of Major Elliott, the 2010 Kunar PRT CAT defined its primary mission as "help[ing] increase the capacity of the host nation government ... [by] effectively employ[ing] CA assets in Kunar Province."[14]

The Kunar PRT, along with USAID, had invested large sums improving the quality of life, infrastructure, stability, and economic opportunities available to Kunar inhabitants. Since 2008, the Kunar PRT built schools, repaired roads, constructed bridges, opened a three story courthouse in Asadabad, erected a new 100 cell provincial prison, and paved the Pech River valley road that links Asadabad with Nuristan Province directly to the north. The PRT also promoted agriculture. In late 2008, for instance, the PRT distributed 300 pounds of sugar to Kunar beekeepers in order to help feed the bees and assist with the spring pollination of crops. The PRT also launched a 30 minute call in agricultural radio show in which Kunar farmers asked questions related to fertilizer application, livestock husbandry, irrigation, and other related topics. Seeking to improve food preservation skills, the PRT's Female Engagement Team hosted workshops in 2008 on fruit and vegetable pickling techniques.[15]

In 2010, the Yargul flooding crisis was one of the incoming PRT CAT's first assignments. Commander Edwards worried that the situation, if left unresolved, promoted discontent and undermined base security. Surrounded by cornfields, wheat fields, and tree lines, the culvert denial grate provided the ideal setting for a Taliban ambush. If insurgents observed the PRT visiting the aqueduct after every rainstorm and performing grate maintenance, they could launch an ambush from concealed positions and inflict heavy casualties. Edwards was unwilling to jeapordize US Soldiers

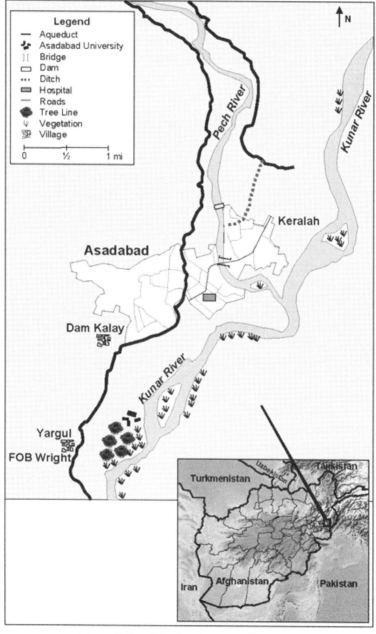

Figure 3. Asadabad and surrounding area.

to clean the grate. Yet he did not want to alienate Yargul villagers by ignoring their grievances.

Edwards instructed the PRT CAT to identify potential solutions.[16] Ladisic took the lead in devising a flood control program. As an armor officer in the Pennsylvania National Guard, Ladisic had little formal experience with civil affairs or Afghan culture before his deployment. In general, National Guard Soldiers received theater briefs on regional culture, ethnicity, language, and religion before deployment. Those interested in obtaining more detailed briefings sometimes requested additional information from the unit they replaced. Yet Ladisic did not have this opportunity. In the weeks leading up to his deployment, he expected to serve in Laghman Province as an operations officer with the 1st Battalion, 178th Field Artillery Regiment (1-178 FA) from the South Carolina National Guard. Confusion and uncertainty filled Ladisic's final days before deployment. At the last minute, the 1-178 FA received a change of mission and was ordered to take over security at the Kabul International Airport. Ladisic's orders also changed. On the day before he departed the United States, he learned that his new assignment was as the CAT project manager for the Kunar PRT. With essentially no preparation for his insertion into Kunar politics and culture, Ladisic knew that he faced a steep learning curve.[17]

Arriving at FOB Wright on 22 February 2010, Lieutenant Ladisic began his duties. Seeking to devise a flood control and garbage disposal solution, he gathered information and consulted with Major Elliott and Sergeant First Class LeMond. Although the PRT CAT understood that the cleanup program was a marginal, temporary, and inexpensive development project, they hoped to forge a friendly relationship with Sher, eliminate the source of grievances between Yargul and the US military, and resolve a thorny environmental problem. In the absence of legitimate and effective local governing institutions, the PRT CAT also intended to inspire Yargul residents to take responsibility for their village's welfare.

Considering a broad range of options, Ladisic walked himself through every potential solution in order to determine the best course of action. His creative, resourceful, and exhaustive consideration of solutions was a testament to his adeptness at adapting to challenging and unfamiliar circumstances in which he had little formal training or instruction.

The first solution he considered was to do nothing. Even though this option was appealing because the culvert denial grate was *outside* of FOB Wright, doing nothing risked estranging the Afghans that lived just outside of the base. Furthermore, Yargul teenagers worked in a variety of roles at FOB Wright, and Sher, a key local powerbroker, was friendly toward coalition forces. If Sher's complaints about the flooding went unanswered, his villagers might question the authority and influence he wielded with the PRT and display hostility towards US Soldiers. Given these grim assumptions, Ladisic concluded that doing nothing constituted an ill-advised and counterproductive strategy.

Ladisic's second option involved hiring Yargul teenagers to keep the grate unobstructed. This plan was attractive because it offered an opportunity to put money into the pockets of Yargul's young men deterring them from joining the insurgency and demonstrating the PRT's solidarity with the small village. Unfortunately, it also presented unwieldy logistical and supervisory issues. Ladisic wondered how the PRT CAT would supervise the cleanups, reimburse the workers, and ensure that no one was injured. Uncertainty about oversight and supervisory responsibilities led him to reject this proposal in favor of other strategies.

Hiring a local contractor constituted the third possible solution. Yet security concerns also made this option unappealing. If the contractors arrived at the grate after every rainstorm, they too might be vulnerable to an ambush. Since contractor safety and security were the PRT's responsibility, employing contractors would divert scarce resources and assets away from other urgent reconstruction and development programs. In addition to safety concerns, Ladisic also frowned on the idea of paying an outside contractor when the money could be used to benefit Yargul residents or politically empower Sher. Small wonder he also dismissed the contractor option.

The final alternative proved most advantageous. After rejecting all of the other options, Ladisic authorized a cash-for-work program. The structure of the program was simple and straightforward. Ladisc would reimburse Sher for hiring Yargul villagers to clean the grate and remove the trash on a periodic basis. There were many benefits to this arrangement. The cash-for-work program promised to reduce flooding, remove debris from the aqueduct, provide an infusion of cash into Yargul, elevate Sher's status, and solidify a political alliance between the PRT CAT and Sher. Due to Yargul's proximity to FOB Wright, establishing solidarity with the village's elder was critical to preserving base security.[18]

On 5 March 2010, Ladisic formally submitted a request for bulk CERP funds in order to initiate the cash-for-work program. As the PRT CAT awaited a funding decision, Ladisic convened a series of meetings at

Sher's home and invited him to lead the cleanup effort. They agreed that the elder would pay his workers seven to nine dollars per day to keep the grate unobstructed. This pay exceeded the Afghanistan national average by three times. Ladisic also approved the purchase of tools, rakes, and wheelbarrows to haul away the trash. The PRT CAT also identified a second cleanup site. The hospital in Asadabad, five miles north of Yargul, carelessly discarded hypodermic needles onto the floodplain of the nearby Kunar River. Since Afghan children enjoyed playing on the river's banks, the contaminated needles represented a major public safety menace. In instructing Sher to remove the needles from the banks of the Kunar River, he intended to eliminate a public health hazard and cultivate goodwill by demonstrating the PRT CAT's commitment to improving the quality of life for Asadabad families.

Six weeks later, on 12 April, the PRT CAT was awarded $2,100 in CERP funds to initiate the program.[19] "Everything we did," Ladisic emphasized, "was...to give people confidence in their leaders and their government."[20] Elevating Sher to a leadership role legitimized his political status and reinforced his right to employ power in his own village. On 16 April, Ladisic invited Sher to FOB Wright. Finalizing the terms of the cash-for-work program, Ladisic told him to return the next day with a list of trustworthy and responsible villagers who he hired to clean the grate, dispose of the trash, and collect the contaminated needles scattered near the Kunar River. He also laid out the PRT CAT's expectations. Allegations of fraud, corruption, or the misuse of funds would not be tolerated. Since the primary purpose of the CERP funds was to empower Sher and enhance his political standing and prestige, Ladisic insisted that he exercise complete autonomy and supervision. The success, or failure, of the cash-for-work program rested on the elder's integrity. As Sher departed FOB Wright, Ladisic felt confident that he had enlisted an elder of unimpeachable integrity and character to administer the undertaking. The cleanup program was set to begin.[21]

The next morning Sher brought a list of names, a cleanup schedule, and an estimate of how much he intended to pay each worker. Satisfied with the elder's preparations, Ladisic again demanded strict accountability. The PRT CAT would evaluate progress by making unannounced visits to the grate and the needle cleanup site. Unsatisfactory results might lead to the program's early termination. Nevertheless, before the meeting concluded, Ladisic gave Sher the full $2,100. In an attempt to foster accountability, the PRT CAT also mandated that he collect the signature or fingerprint of every worker each time he paid them.[22]

The Clean Up Commences

Sher tackled his new responsibilities with enthusiasm and eagerness. On 18 April 2010, he organized the first culvert denial grate cleanup. Using rakes, wheelbarrows, and other tools purchased with bulk CERP funds, the Yargul villagers, mostly teenage boys aged 12-19, removed trash and other debris from the grate.[23] After the initial grate cleanup, the villagers proceeded to the Asadabad hospital. Under Sher's direction, they picked up used hypodermic needles from the banks of the Kunar River. Afterward, he reimbursed his workers the predetermined rate, returned to FOB Wright, and submitted a receipt containing the signatures and fingerprints of paid workers.[24]

The cash-for-work program yielded immediate results. By May, rain runoff and snowmelt from the Pech River watershed flowed through the aqueduct unimpeded. For the first time since the implementation of the culvert grate denial program, Yargul homes, crop fields, and roads were not submerged after downpours.[25] The program initially attained its goals for the local economic impact, Sher's status, and his relationship with coalition forces. According to Ladisic, Sher was "appreciative of the PRT CAT supporting his village and empowering him in front of his villagers."[26]

Photo Courtesy of First Lieutenant Joseph Ladisic

Figure 4. Yargul villagers collecting needles on the banks of the Kunar River.

Despite the initial successes, the cash-for-work program soon triggered tension between Ladisic and Sher. During a routine patrol to check on the culvert denial grate in late May, Ladisic discovered Sher's workers sitting idly on a nearby retaining wall even though garbage covered the grate. The discovery irritated Ladisic. It demonstrated that the villagers, without strict supervision, were willing to take advantage of American generosity. Hoping to persuade Sher to adopt a more assertive supervisory role, Ladisic visited the elder's home. Employing a combination of praise and criticism, the lieutenant commended Sher for keeping the grate unobstructed for five weeks yet he emphasized that he would no longer tolerate lazy or unaccountable workers. He asked Sher to personally visit or send his financial adviser to check on the workers every day that they were dispatched to clean the grate. The tense encounter had the desired outcome. Receptive to criticism, Sher became more active in ensuring that workers did not take advantage of the program.[27]

Ladisic's hands-on leadership also improved the results of the hypodermic needle cleanup. On 18 April, Sher and the Yargul villagers had launched the needle cleanup outside of the Asadabad hospital. Three weeks later during a scheduled patrol, Ladisic discovered that only half of the needles had been picked up. Discarded hypodermic needles littered the banks of the Kunar River where children played and families retrieved water. Disappointed but undeterred, Ladisic instructed to Sher accompany him to the hospital the next day. As the pair walked along the beach, the Yargul elder agreed to have his workers return and complete the cleanup. A few weeks later, Ladisic discovered the beach completely devoid of trash, hypodermic needles, and other medical garbage.[28] Afghan children were finally free to play near the Kunar River without the threat of injury by contaminated needles.

Ladisic next attempted to secure the permanent success of the needle cleanup. After Sher's workers removed the final batch of needles from the beach, he visited the hospital and requested a meeting with Dr. Fazli, the hospital director. Once the two men exchanged customary greetings, Ladisic explained the purpose of the PRT CAT's cleanup project, emphasized the dangers of not properly disposing used needles, and asked Dr. Fazli to eliminate the reckless dumping of medical garbage near the river. Unaware of the PRT CAT's cleanup program or the needle disposal problem, Dr. Fazli altered the hospital's disposal method. Ladisic was pleased by Fazli's demeanor, response, and positive attitude. The beach near the Kunar River remained free of needles for the remainder of Ladisic's deployment.[29]

Unintended Consequences

As the cash-for-work program entered its second month in mid-May, major unanticipated difficulties developed. Faiz Muhammad Khan, a tribal elder from Dam Kalay, an Asadabad suburb north of Yargul, visited the Kunar governor's compound.[30] Hostile and animated, Faiz angrily accused the PRT of subverting his political and social authority by selecting Sher to head the cleanup program. Yargul village, he explained, was part of the Dam Kalay grand *shura*. In Afghan tribal culture, *shuras* are temporary advisory councils that mediate local conflicts and exercise civic responsibilities for law and order, justice, security, humanitarian needs, and reconstruction. Faiz argued that the PRT CAT destabilized the local political hierarchy by appointing Sher, rather than himself or one of his family members, to head the cash-for-work program.[31]

Faiz's criticism caught the Kunar PRT CAT off guard. During Ladisic's organizational meetings with Sher, the Yargul elder never mentioned his village's subordinate social relationship to Dam Kalay. Nevertheless, the brewing controversy demanded a swift and amicable resolution. By intensifying tribal rivalries and tensions, the program was becoming counterproductive. Looking to defuse the tense situation, Ladisic met on several occasions with Faiz. During the discussions, the Dam Kalay elder stated how he would have used the CERP funds to enrich his family. Ladisic was stunned. On the one hand, the lieutenant admitted that the PRT CAT exercised poor judgment by selecting Sher rather than Faiz to lead the program. Yet Ladisic also concluded that Faiz was only interested in supervising the cleanup to line his own pockets. He wanted something for nothing. Taking a firm stance, Ladisic admonished Faiz. He emphasized that the purpose of cash-for-work was to economically empower the entire community of villages and not his family. "Look," Ladisic instructed him, "that's not what these programs are about. It's not about helping your family. If you don't understand that, I'm sorry but it's about helping everybody."[32]

Despite Ladisic's stern rebuke, Faiz escalated the conflict. Over the next month he visited the governor's compound on several occasions claiming that he also served as the local "ditch and canal director." Ratcheting up his rhetoric, he angrily denounced the PRT CAT and demanded that Ladisic turn control of the cleanup program over to his family. The episode revealed the significance of cultural awareness in all US Army operations in Afghanistan. Navigating Afghanistan's tangled maze of competing political factions and rival tribes was a prerequisite for forging meaningful political alliances with relevant local powerbrokers.

The cash-for-work program, rather than building harmonious political alliances between coalition forces and village elders in the vicinity of Asadabad, instead provoked intertribal discord, tension, and jealousy. The controversy undermined the PRT CAT's relationship with communities in the volatile Asadabad area.[33]

The PRT CAT eventually decided to mollify Faiz. In July, Ladisic, First Lieutenant Vinodhini Darmarajah, and other CAT members approached the Dam Kalay village elder about initiating a CERP project under his sponsorship in Asadabad. They agreed to build a railing on a steep set of stairs behind the provincial governor's compound. According to Darmarajah, the $1,500-project promoted safety, provided jobs for Asadabad and Dam Kalay residents, demonstrated that the PRT did not favor Yargul at the expense of other communities, and helped "smooth things over" with Faiz.[34] The CERP funded railing project took eight weeks to complete and employed local workers. Darmarajah identified the railing project as a key reason why the tension between the PRT CAT and Faiz slowly subsided over the course of the summer. Cash, if properly used, proved an effective tool to solve problems and reconcile feuding parties.[35]

As the cleanup program entered its final month in mid-June, additional cultural misunderstandings hampered success. In launching the cleanup program, the PRT CAT anticipated that Sher's workers would collect, remove, and dispose the garbage that accumulated on the culvert denial grate. This assumption was based on the faulty premise that Afghans, like western people, identified trash as an expendable and disposable material. Rather than hauling trash away from the aqueduct, however, Sher's workers simply moved it to the area adjacent to the grate. Heavy downpours washed the garbage back onto the culvert denial grate. When questioned by Ladisic about the worker's strategy, Sher responded that the landfill concept was alien to Kunar Province. Afghans reused trash as fuel and building supplies. According to Sher, it was foolish, shortsighted, and irresponsible to bury garbage so that it could never be salvaged or burned as a fuel source. His villagers would keep the grate unobstructed but they did not intend to transport the trash away from the aqueduct.[36]

On 17 July, the cleanup program ended. Over the next several weeks, the PRT CAT watched with curiosity whether the cash-for-work program fostered a sense of community responsibility in Yargul about flood control and garbage removal. Unfortunately, when the bulk CERP funds dried up, Sher and his villagers lost interest in the aqueduct. As soon as the rains returned, so too did the floods.

Aftermath

The Kunar PRT CAT culvert grate denial cleanup program exemplified the daunting challenges faced by US Army Soldiers in navigating Afghanistan's complex labyrinth of local political hierarchies, cultural traditions, and competing powerbrokers. One anthropologist described the fragmented nature of Afghanistan's social and political culture by stating that "the outstanding feature of social life in Afghanistan is its local tribal or ethnic divisions. People's primary loyalty is respectively to their kin, village, tribe, or ethnic group."[37] Incoming PRT leaders had to obtain a comprehensive awareness of Afghan politics, culture, and society in order to efficiently operate in this difficult environment. In a fragmented society, cultural awareness was a prerequisite for forging and sustaining productive alliances with local and provincial powerbrokers. The Kunar PRT CAT cleanup program proved counterproductive because team leaders were unaware of the long standing social and cultural relationships that defined everyday life and the interaction of villages in Asadabad's hinterland. In hindsight, the PRT CAT's assumption that money alone could alter longstanding cultural understandings of waste was shortsighted.

Ultimately, feelings of disillusionment and bitterness overwhelmed the PRT CAT. Despite the US Army's noble and well intentioned efforts, the cleanup and flood control program failed to inspire Yargul villagers to seize the initiative and take responsibility for their own welfare. "[The villagers were] not going to go above and beyond unless they're getting paid for it," Ladisic concluded.[38] Elliott shared his subordinate's sentiment. Compared to other CERP projects authorized by the Kunar PRT CAT, the cleanup program, in Elliott's words, was a "sideshow," a "drop in the bucket," and "a waste of time."[39] The program's major shortcoming was that it attacked the symptom of the flooding problem (the clogged grate) rather than the ultimate source (Afghan cultures' proclivity towards discarding trash in waterways). Nevertheless, the PRT CAT's responsiveness to unanticipated and adverse events such as the chronic garbage problem and intertribal friction was a testament to creative, proactive, and engaged leadership by company grade officers.

The program did lead to notable successes and promoted a stable security environment. During the past decade, a growing body of scholarship argued that contemporary insurgencies sometimes sabotage critical energy and water infrastructure in order to destabilize and undermine the credibility of local governments.[40] A US Army study argued that during Operation IRAQI FREEDOM, the insurgency based in Sadr City lost momentum once basic sanitation, electrical, and water services were restored and protected.[41] Elliott contended that the cash-for-work program, by reducing floods attributed to the American construction project, probably promoted base security by deterring Yargul villagers from sympathizing with the Taliban. Throughout 2010, no major security incidences occurred at Yargul. The PRT discovered that it was difficult to enhance the authority and prestige of national governing institutions in such a parochial and fragmented society but Elliott applauded his PRT CAT's record of promoting base stability. Security concerns always trumped the flooding and trash problems. "The effect that we were trying to achieve had nothing to do with the clean grate. Ultimately, what we wanted was to make sure that the people that lived next to the wall weren't going to throw a grenade over...So, did the project fail? Well, we didn't solve the trash problems...but did it achieve the effect in that we created some goodwill with the Yargul villagers that lived next door to us? Yeah, I think so and that, in my mind, is probably good enough."[42]

Notes

1. First Lieutenant Joseph Ladisic, interview by Anthony E. Carlson, Combat Studies Institute, Fort Leavenworth, KS, 24 June 2011, 10-11.

2. Second Lieutenant Amy Abbott, "Cash for Work program employs local Afghan villagers," 24 April 2010, http://troopscoop.posterous.com/daily-afghan-iraq-update4-24 (accessed 21 June 2011).

3. Ladisic, interview, 10.

4. Office of the Special Inspector General for Afghanistan Reconstruction, "Commander's Emergency Response Program in Laghman Province Provided Some Benefits, but Oversight Weaknesses and Sustainment Concerns Led to Questionable Outcomes and Potential Waste," www.sigar.mil/pdf/audits/SIGARAudit-11-7.pdf (accessed 1 August 2011), ii, 1.

5. Major Brian P. Elliott, interview by Anthony E. Carlson, Combat Studies Institute, Fort Leavenworth, KS, 30 June 2011, 20, 16.

6. *Afghanistan Provincial Reconstruction Team: Observations, Insights, and Lessons* (Ft. Leavenworth: Center for Army Lessons Learned, 2011), 93-4; Michael Moore and James Fussell, "Kunar and Nuristan: Rethinking US Counterinsurgency Operations," *Institute for the Study of War*, Afghanistan Report 1 (July 2009): 6-7.

7. Moore and Fussell, 20; Christopher Bodeen, "US Troops Withdraw from Korengal Valley," *Army Times*, 14 April 2010, www.armytimes.com/news/2010/04/ap_afghanistan_korengal_041410/ (accessed 8 August 2011).

8. J. Chivers, Alissa J. Rubin, and Wesley Morgan, "US Pulling Back in Afghan Valley it Called Vital to War," *New York Times,* 24 February 2011, www.nytimes.com/2011/02/25/world/asia/25afghanistan.html (accessed 8 August 2011).

9. Donald P. Wright, James R. Bird, Steven E. Clay, Peter W. Connors, Lieutenant Colonel (LTC) Scott C. Farquhar, Lynne Chandler Garcia, and Dennis F. Van Wey, *A Different Kind of War: The United States Army in Operation ENDURING FREEDOM (OEF), October 2001-September 2005* (Fort Leavenworth: Combat Studies Institute, 2010), 3, 254-55.

10. *Afghanistan Provincial Reconstruction Team*, 1-3; Moses T. Ruiz, "Sharpening the Spear: The United States' Provincial Reconstruction Teams in Afghanistan," *Applied Research Projects, Texas State University-San Marcos* 297 (Spring 2009), 57.

11. Kenneth Katzman, "Afghanistan: Post-Taliban Governance, Security, and US Policy," (n .p.: Congressional Research Services, 2010), 49-50; Katzman, "Afghanistan: Post-Taliban Governance, Security, and US Policy," (n. p.: Congressional Research Services, 2011), 27-8; *Provincial Reconstruction Teams: An Interagency Assessment* (n. p.: USAID, 2006), 8-9; Wright et. al., 254-59.

12. *Afghanistan Provincial Reconstruction Team*, 3.

13. North Atlantic Treaty Organization, "Provincial Reconstruction Teams (PRTs)," 8 November 2010, http://www.nato.int/isaf/topics/prt/ (accessed 22 August 2011).

14. For an organizational chart of the Kunar PRT CAT, see Elliott, First Lieutenant Vinodhini Darmarajah, Ladisic, Master Sergeant Donald David, Sergeant First Class Christina LeMond, Sergeant Anthony Didonato, Corporal Toan Q. Pham, and Tony Unite, "Provincial Reconstruction Team Kunar: Civil Affairs Best Practices," unpublished paper, 2010, 7. See also Elliott, e-mail to Anthony E. Carlson, Combat Studies Institute, Fort Leavenworth, KS, 6 September 2011.

15. Lily J. Lawrence, "Road Improves Life for Pech Valley Communities," 17 July 2008, *Central Asia Online*, http://centralasiaonline.com/cocoon/caii/xhtml/en_GB/features/caii/features/2008/07/17/feature-03 (accessed 7 September 2011); Second Lieutenant Neil Myers, "PRT Supports Agriculture in Konar Province," 16 August 2008, Defense Video and Imagery Distribution System, http://www.dvidshub.net/news/printable/22571 (accessed 7 September 2011); Soraya Sarhaddi Nelson,"Westerners Play Pivotal Role in Afghan Rebuilding," *National Public Radio*, 20 May 2008, http://www.npr.org/templates/story/story.php?storyId=90599416 (accessed 7 September 2011); Lieutenant Junior Grade James Dietle, "Afghanistan Department of Agriculture, PRT Work to Expand Beekeeping in Konar Province," 1 January 2009, Regional Command-East, http://www.rc-east.com/en/regional-command-east-news-mainmenu-401/1530-afghanistan-department-of-agriculture-prt-work-to-expand-beekeeping-in-konar-province.html (accessed 7 September 2011); "Kunar PRT Attends Prison Opening," 23 December 2010, Defense Video and Imagery Distribution System, http://www.dvidshub.net/news/printable/62713 (accessed 7 September 2011); "Kunar PRT Helps Young Women Learn Food Preservation Skills," 31 May 2011, Defense Video and Imagery Distribution System, http://www.dvidshub.net/news/printable/71331 (accessed 7 September 2011); First Lieutenant Nicholas Mercurio, "US Ambassador, PRT Inaugurate New Courthouse in Kunar," 13 May 2011, Regional Command-East, http://www.rc-east.com/ar/press-releases-mainmenu-326/4502-us-ambassador-prt-inaugurate-new-courthouse-in-kunar.html (accessed 7 September 2011).

16. Ladisic, interview, 10-11.

17. Ladisic, interview, 39-40; and Ladisic, e-mail to Anthony E. Carlson, Combat Studies Institute, Fort Leavenworth, KS, 22 August 2011.

18. For a description of all of the strategies Ladisic and the PRT CAT considered, see Ladisic, interview, 14-5.

19. Ladisic, interview, 12-3, 15, 36.

20. Ladisic, interview, 16.

21. Ladisic, interview, 11-13.

22. Ladisic, interview, 11-13.

23. Ladisic, e-mail to Anthony E. Carlson, Combat Studies Institute, Fort Leavenworth, KS, 15 August 2011.

24. Ladisic, interview, 16-18.

25. Ladisic, interview, 18.

26. Ladisic, interview, 31.

27. Ladisic, interview, 18.

28. Ladisic, e-mail, 15 August 2011.

29. Ladisic, interview, 35.

30. The name of the Dam Kalay elder was given in Darmarajah, e-mail to Anthony E. Carlson, Combat Studies Institute, Fort Leavenworth, KS, 18 August 2011.

31. Ladisic, interview, 27. On the history of Afghan shuras, see Lynn Carter and Kerry Connor, *A Preliminary Investigation of Contemporary Afghan Councils* (Peshwar, Pakistan: Agency Coordinating Body for Afghan Relief, 1989), 2-10; and Ali Wardak, "*Jirga* – A Traditional Mechanism of Conflict Resolution in Afghanistan," United Nations Online Network in Public Administration and Finance, 2003, unpan1.un.org/intradoc/groups/public/.../apcity/unpan917434.pdf (accessed 17 August 2011).

32. Ladisic, interview, 29-30.

33. Ladisic, interview, 30; Ladisic, e-mail to Anthony E. Carlson, Combat Studies Institute, Ft. Leavenworth, KS, 12 July 2011.

34. Darmarajah, e-mail to Anthony E. Carlson, Combat Studies Institute, Ft. Leavenworth, KS, 26 July 2011.

35. Darmarajah, e-mail, 26 July 2011; and Darmarajah, interview by Anthony E. Carlson, Combat Studies Institute, Fort Leavenworth, KS, 29 June 2011, 14-15.

36. Ladisic, interview, 33.

37. Thomas Barfield, *Afghanistan: A Cultural and Political History* (Princeton: Princeton University Press, 2010), 18.

38. Ladisic, interview, 41.

39. Elliott, interview, 16-22.

40. Chad M. Briggs, "Environmental Change, Strategic Foresight, and Impacts on Military Power," *Parameters* 40 (Autumn 2010): 81-82.

41. Colonel Timothy E. Hill, *Reducing an Insurgency's Foothold: Using Army Sustainability Concepts as a Tool of Security Cooperation for AFRICOM* (Arlington: Army Environmental Policy Institute, 2008), 11-14. On the importance of water in conducting counterinsurgency operations, see Department of the Army, *Field Manual (FM) 3-24, Counterinsurgency* (Headquarters, Department of the Army: Washington, DC: 2006), 1-9, 2-2, 3-11, 5-23, 6-2.

42. Elliott, interview, 20-21.

Operation STRONG EAGLE

Combat Action in the Ghakhi Valley

by

John J. McGrath

In the summer of 2010, Coalition forces in eastern Afghanistan faced a mounting insurgent threat in districts along the border with Pakistan. In order to contest this rising menace, in June 2010 the 2d Battalion, 327th Infantry (2-327th IN) (Task Force (TF) *No Slack*), conducted Operation STRONG EAGLE in the Marawara District of Kunar Province. The goal of the operation, the first major offensive conducted by the newly arrived 1st Brigade Combat Team (BCT), 101st Airborne Division (TF *Bastogne*), was to reduce enemy presence in the heavily-populated Kunar River Valley region of Kunar Province.

At the time of the operation, General David Petraeus was awaiting Senate confirmation of his appointment to the post as head of the Afghanistan Theater of operations based out of Kabul. Petraeus would assume command of the International Security Assistance Force (ISAF) several days after the conclusion of STRONG EAGLE replacing General Stanley McChrystal. ISAF was a NATO post. As the senior American officer in Afghanistan, Petraeus also headed the US Forces in Afghanistan (USFOR-A) command, a headquarters established in 2008 to oversee the activities of all American troops in the theater. Under Petraeus was the International Joint Command (IJC), led by Lieutenant General David M. Rodriguez since its establishment in 2009. Rodriguez, whose headquarters was at the Kabul airport, controlled ISAF's operational forces. In 2010 these included six regional commands. Three of the regional commands were led by and predominately staffed with American forces. One of these headquarters, Regional Command East (RC-East) based at Bagram Airfield, controlled operational units in northeastern Afghanistan. In June 2010, Headquarters, 101st Airborne Division, called *Combined Joint Task Force 101* (CJTF-101) while deployed and commanded by Major General John Campbell, functioned as the RC-East Command. The division controlled its own 1st and 3d BCTs based at Jalalabad and Khowst respectively along with the 173d Airborne BCT (Logar), the Vermont Army National Guard's 86th BCT (Bagram), a French brigade, and a Polish brigade. The 1st BCT, 101st Airborne Division (TF *Bastogne*), located at the Jalalabad Airfield (Forward Operating Base [FOB] Fenty) and with Colonel Andrew Poppas commanding, was responsible for the four northeastern provinces of Nangahar, Laghman, Nuristan and Kunar. Poppas controlled four maneuver battalions. One of these units, the 2d Battalion, 327th Infantry, commanded by Lieutenant Colonel Joel Vowell, had responsibility for the eastern portion of Kunar Province, including the southern portion of the Kunar River Valley and Marawara District.[1]

Special operations forces (SOF), composed of a mix of US Army Special Forces teams, Navy SEALS, and other SOF elements functioned under a separate chain of command directly under Petraeus. US Army Brigadier General Austin Miller commanded the Combined Forces Special Operations Component Command (CFSOCC) at Kabul. Under Miller was Colonel Donald Bolduc's Combined Joint Special Operations Task Force (CJSOTF). Bolduc oversaw four regionally based battalion-sized SOF task forces. One of these task forces had responsibility for the area covered by RC-East. In addition to conducting special operations in RC-East, the SOF were the advisors for several elite Afghan commando and other units that often fought alongside American troops and non-elite elements of the Afghan National Security Forces (ANSF).[2]

Background

STRONG EAGLE was the culmination of a long series of operations in Kunar Province since the beginning of Operation ENDURING FREEDOM (OEF) in 2001. Coalition presence was minimal in northeastern Afghanistan for the first two years of OEF. The Taliban forces in the area withdrew under minimal local pressure in November 2001 concurrent with their withdrawal from the city of Jalalabad in Nangarhar Province just to the south of Kunar. Subsequently, US forces obtained control of a former Soviet and Afghan military compound south of Asadabad and established a small firebase there to coordinate the future activities of special operations forces (SOF) in the region. This outpost, originally called simply Firebase Asadabad and later Camp Wright, supported a company of paratroopers from the 82d Airborne Division in 2002. Special Forces elements also established another outpost farther north along the Kunar River at Naray, a post later redesignated Forward Operating Base Bostick. From these outposts, SOF elements initially conducted operations to engage the local populace, primarily through the conduct of medical civic assistance programs (MEDCAP) and missions designed to capture high value targets (HVT) enemy leadership cells. These types of operations predominated in the region through mid-2006, even after the introduction of conventional forces which either augmented the activities of the SOF units or operated temporarily in the region to conduct a specific operation.[3]

Figure 1. Marawara District in Relation to the Rest of Afghanistan.

In 2003 and 2004 light infantry forces from the 10th Mountain Division and later the 82d Airborne Division provided small security elements for the Kunar Valley bases at Asadabad and Barikowt north of Naray on the Pakistani border. When intelligence sources indicated that important insurgent leaders were meeting in the Waygal Valley region of Nuristan, just north of Kunar, US forces executed Operation MOUNTAIN RESOLVE in early November 2003. MOUNTAIN RESOLVE involved elements of two infantry battalions from the 10th Mountain Division, Army Rangers and Special Forces operators and Navy SEALs. In the operation, US troops were inserted by helicopter into the Waygal Valley and conducted area sweeps up to the town of Aranas in central Nuristan Province. The most enduring result of MOUNTAIN RESOLVE was the creation of a new firebase near the village of Nangalam in the Pech Valley. This site later became known as Camp Blessing.[4]

After MOUNTAIN RESOLVE, the number of conventional units in Kunar Province gradually increased. From November 2003 to June 2006, a series of US Marine infantry battalions on seven-month rotations stationed elements in Kunar Province. Eventually the Marines maintained a company at Asadabad with one of its platoons at Camp Blessing. SOF operations continued in the area and the Marines at times massed forces for particular operations. The most significant of these efforts were Operations RED WINGS and WHALERS in 2005 in which the 2d Battalion, 3d Marine Regiment (2/3d Marines), sent converging columns into the Korengal valley south of the Pech River.[5]

Conventional Army forces moved into Kunar Province in large numbers for the first time beginning in April 2006 as a result of Operation MOUNTAIN LION. Elements of two different battalions (1-32d IN and 3d Squadron, 71st Cavalry) from the 3d Brigade Combat Team, 10th Mountain Division along with part of the 1st Battalion, 3d Marines (1/3d Marine Regiment) again converged on the Korengal Valley. However, at the conclusion of operations there, as the Marines departed for their home station in Hawaii, the Army elements then moved to permanent garrison locations in Kunar and Nuristan provinces, establishing a series of small outposts along the major river valleys. From June 2006 to January 2009, this two-battalion force, through three unit rotations, conducted counterinsurgency operations throughout Kunar Province. One battalion had responsibility for all but the two northernmost districts of Kunar and for the three central districts of Nuristan Province. The other, a series of cavalry squadrons, had responsibility for the two northernmost Kunar districts and the two easternmost districts of Nuristan Province.

In order to increase the coalition presence along the Pakistani border area, commanders in the Coalition added another maneuver battalion to the forces in northeastern Afghanistan (the RC-East area) in January 2009. To provide this reinforcement, the incoming 1st Battalion, 32d Infantry (1-32d IN) was detached from its parent brigade and assigned to the 3d Brigade Combat Team, 1st Infantry Division, the unit responsible for Kunar Province. The new battalion had previously served in Kunar in 2006 and 2007. With the 1-32d IN's arrival, the sector formerly the responsibility of a single battalion (in January 2009 the 1st Battalion, 26th Infantry) was divided in half with the new unit assuming responsibility for the ten Kunar districts closest to Pakistan. Further complicating things was the fact that 1-32d IN was on a different rotation cycle from the other units in Kunar. Accordingly, midway through the battalion's one year tour, all the other units in the province returned to their home stations and the 4th BCT, 4th Infantry Division, replaced the 3d BCT, 1st Infantry Division as the responsible brigade headquarters. In late 2009, 1-32d IN itself departed Afghanistan being replaced in Kunar, (for the second consecutive time in OEF) by the 2d Battalion, 503d Infantry (2-503d IN). The 2-503d IN was detached from its parent unit, the 173d Airborne Brigade Combat Team to the 4th BCT, 4th Infantry Division. When the 1st BCT, 101st Airborne Division, replaced the 4th BCT in Kunar in early 2010, Major General Campbell, the 101st commander, preferred to retain all his organic maneuver battalions under his control for the unit's full tour. Therefore the 2-503d IN was sent back to its parent brigade and replaced in eastern Kunar by the 2d Battalion, 327th Infantry (Task Force *No Slack*). Operation STRONG EAGLE was the new battalion's first major operation upon assuming responsibility for this region.[6]

Per one of the components of US Army counterinsurgency doctrine, the primary mission of all these eastern Kunar battalions was to interdict enemy movement and lines of communication from across the border out of Pakistan. This was chiefly accomplished through the posting of troops in outposts centered on population centers and conducting routine and continuous patrols throughout specific company sectors. Occasionally the battalions would mass forces for larger operations and periodically, troops were dispatched to areas of increased insurgent activity out of sector such as Barg-i-Matal, in Nuristan in 2009 and 2010.[7]

The Ghakhi Valley

The Kunar River Valley runs northeast from the Kabul River Valley near Jalalabad for about 60 miles through the Afghan provinces of Nangarhar and Kunar before flowing into Pakistan. Within Kunar Province, the river's course flows parallel to and roughly 10 kilometers (five miles) west of the Pakistani border (the famous Durand Line). The border is demarcated by the peaks of the Hindu Raj range, a southern extension of the Hindu Kush. There are numerous passes through this ridgeline. The one closest to the provincial capital of Asadabad is the Ghakhi Pass in Marawara District.

Marawara District is located directly across the Kunar River from and to the east of Asadabad, the largest town in the province. The main geographical feature of the district is the Ghakhi Valley (Figure 2), which extends for roughly 11 kilometers (six miles) from the district capital, Marawara village on the Kunar River, to the Ghakhi Pass located on the Pakistani border. A newly built bridge across the Kunar River connected Marawara village with the Kunar Valley road running north from Asadabad on the west side of the river. A major unpaved road ran from Marawara to the Ghakhi Pass through the villages of Sangam, Daridam, and Chinar. Intelligence analysts believed that Chinar, the most easterly of these settlements, was the headquarters of the Taliban commander for Kunar.

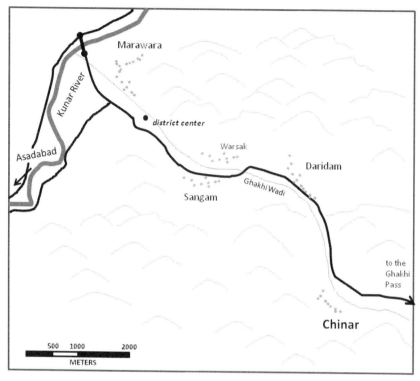

Figure 2. The Ghakhi Valley.

Marawara District's population consisted entirely of *Mohmand Pashtuns*, who also live in nearby Pakistan and had settle the region east of the Kunar River in the early nineteenth century.[8]

Coalition forces had long conducted humanitarian and development projects in Marawara District. In 2008 the Kunar Provincial Reconstruction Team (PRT), a US run Coalition agency for economic development, issued a contract to build the previously mentioned bridge across the Kunar River near the village of Marawara. This small bridge, once it was completed in November 2009, allowed the local inhabitants access to the markets and bazaars of Asadabad. Prior to this, most residents crossed the Ghakhi Pass into Pakistan for such resources. Economic aid to the district continued in 2009 and 2010. A new district center was built and an Army National Guard Agribusiness Team frequently visited the district to assist local farmers. Since 2004, the National Solidarity Project (NSP), an Afghan governmental development agency, had built a series of surface level aqueducts along the floor of the Ghakhi Valley to facilitate irrigation. The Kunar PRT also assisted in rebuilding these watercourses in 2008 and

2009. The aqueducts were a boon to local agriculture but also provided natural fighting positions alongside the road.[9]

Mountainous ridgelines dominated the Ghakhi Valley road to both the north and south. The valley floor was at an elevation of slightly less than 1,000 meters, or six-tenths of a mile above sea level. That was the low ground. The surrounding peaks rose to heights as high as 1,800 meters (over a mile above sea level). A *wadi* through which flowed the intermittent Ghakhi stream formed a gorge that paralleled the road in the lowest land in the valley. Past Sangam, the west-east valley road paralleled the *wadi* to the south, eventually crossing it between Sangam and Daridam (Figure 3). Coming from the west, the road angled to a northeasterly direction and plunged down 50 meters into the *wadi* fording the usually dry streambed of the Ghakhi creek. Continuing to the northeast, the road then reclimbed the 50 meters to the valley floor before turning east again to travel to Daridam. In the 1980s, the Soviets called this area the Marawara Pass as the road was particularly dominated by high ground on all sides as it crossed the *wadi*. In the valley, the inhabitants conducted agricultural activities based on artificial terraces, a characteristic feature of northeastern Afghanistan. Just west of Daridam there were pomegranate orchards built on such terraces both to the north and south of the road.

During their campaigns in Afghanistan, Soviet forces conducted military operations in Marawara District in order to close insurgent access to sanctuary areas in nearby Pakistan. Large Soviet forces conducted forays into the Ghakhi valley in May 1983 and in April 1985. The latter action was more significant. On 20 April 1985 the 334th *Spetsnaz* Battalion based at Asadabad, conducted a battalion-sized operation to clear the Ghakhi Valley. One *spetsnaz* company advanced down the Ghakhi Valley road while two other companies covered its movement from the high ground to the north and south of the valley. The *mujahideen* managed to advance down the large *wadi* in the center of the valley behind the Soviets from the west and surround the company advancing along the road in separate groups in the villages of Marawara, Daridam, and Sangam. The commander of the company covering the movement from high ground to the north of the road observed this movement but thought the enemy forces were friendly reinforcements and held fire until it was too late. The Soviet command sent a column of BMP armored fighting vehicles and tanks from the 2d Battalion, 66th Separate Motorized Rifle Brigade to relieve the trapped *Spetsnazi*. This column advanced along the Ghazi valley road eastward until a radio-controlled mine emplaced at the point where the road cut

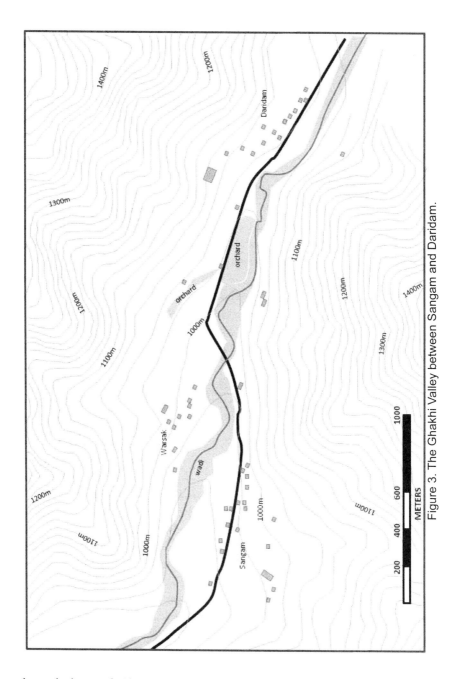

Figure 3. The Ghakhi Valley between Sangam and Daridam.

through the *wadi* (the Maravar or Marawara Pass) disabled the lead tank and stalled the whole relief column for over four hours.[10]

During this delay, the *mujahideen* were able to destroy over half of the trapped *Spetsnaz* elements. The company commander, Captain Nikolai Tsebruk, was killed in Daridam. Eventually the reinforcements were able to rescue the survivors and evacuate the wounded. The Soviets suffered 31 men killed in action and over 100 wounded. For its heroic defense, the 1st Company, 334th *Spetsnaz* Battalion, thereafter carried the informal honorific title of "Marawara Company." Several months later, the *Spetsnaz* battalion commander, feeling the *mujahideen* had used sanctuary areas in Pakistan to assemble for the fight at the Marawara Pass and had received material, if not direct aid from the Pakistanis, conducted an unsanctioned cross border incursion south of Marawara District. While the incursion was successful, the commander was relieved of his post.[11]

Years later, during Operation ENDURING FREEDOM, Marawara's location astride the Pakistani border continued to affect conditions in the district. When the Pakistani military conducted operations against insurgent sanctuary areas in the Bajaur Agency district opposite Kunar Province in the summer of 2008, thousands of refugees fled to Afghanistan. In Marawara, 748 families sought refuge from the fighting in Pakistan. Most stayed with relatives in the Afghan district. Marawara remained a center of conflict in 2009 and early 2010. In November 2009, the insurgents attempted to overrun the Marawara District Center, just east of the village of Marawara. The district center also served as the district police station. The 2-503d Infantry conducted several operations in the Ghakhi Valley in March and April 2010. While these operations reportedly killed a lot of insurgents, the Americans were unable to eject the enemy permanently from the valley or provide a long term presence there.[12]

By 2010, most intelligence sources recognized Qari Zia Rahman (QZR) as the Taliban military commander for Kunar Province. Rahman's base of operations was the Ghakhi Valley. He was a *Mohmand Pashtun* from the village of Barawolo Kalay located in Marawara District south of the valley on the Pakistani border. Rahman, in his mid-thirties in 2010, had operated on both sides of the Pakistani-Afghan border for a number of years. He claimed connections to Arab Islamic scholars and al-Qaeda. In January 2010, press reports indicated that a Predator strike had killed Rahman in Pakistan's Mohmand Agency. However, Rahman soon reemerged in public.[13]

Rahman controlled a force of an estimated 300 fighters in the Ghakhi Valley. Some of these forces were trainees from other areas as QZR had a reputation as a master teacher of insurgents. By 2010, Qari Zia Rahman and his supporters dominated the central and eastern portions of Marawara District. This dominance was very conspicuous. The village elder of Chinar told an American commander, "I am Taliban. My government is the Taliban. Don't ask me to support you because we're with the Taliban."[14] Any combined American/Afghan operation in the Ghakhi Valley could be expected to face fierce resistance. Previous forays there had been opposed tooth and nail. Most American observers expected that any future operation in the valley would be fiercely resisted by up to 300 of Rahman's fighters.[15]

While Marawara remained Rahman's stronghold, there were indications that he was preparing to expand his sphere of influence. In May and June 2010, daily reports indicated that the Marawara District Center was about to be attacked by a large Taliban force. On 7 June, five American Soldiers were killed by an improvised explosive device (IED) north of Asadabad. On 21 June, at a checkpoint along the Kunar River several miles northwest of Marawara, a female suicide bomber killed two more Americans. Coalition intelligence credited Rahman with these attacks as well as the operation of a terrorist cell that fabricated and distributed suicide bomb vests. The latter group was subsequently eliminated by the Afghan National Directorate of Security (NDS).[16]

The 2d Battalion, 327th Infantry, also known during its deployment as TF *No Slack*, part of the 1st BCT, 101st Airborne Division (TF *Bastogne*) and commanded by Lieutenant Colonel Joel Vowell, arrived in Kunar Province in May 2010. TF *No Slack* replaced the 2-503d Infantry, a battalion which then shifted to Wardak Province and reverted to the control of its parent 173d Airborne Brigade Combat Team. TF *No Slack* had responsibility for eastern Kunar Province, including the Kunar River Valley as far north as Asmar, the Chowkay Valley in the southwest and the border passes into Pakistan. The battalion headquarters was located at Camp Joyce outside of Sarkani, on the east side of the Kunar River about eight kilometers south of Asadabad.

Vowell immediately assessed the importance of Marawara District and planned to do something about the enemy threat there as soon as possible. Intelligence reports indicated that the Ghakhi Valley had been an insurgent stronghold for over two years. Vowell viewed such a sanctuary area so close to Asadabad and the Kunar Valley as unacceptable. "This tumor was big. This tumor had the potential of killing the entire province."[17] He intended to strike at Rahman's base as his first major action upon assuming responsibility for the eastern Kunar area. "Nothing like a battalion operation has been done [in the Ghakhi Valley]."[18] As early as 1 June, only days after the battalion assumed responsibility for its sector, Vowell received approval to plan the operation from his brigade commander, Colonel Poppas. Vowell originally scheduled the operation, codenamed STRONG EAGLE, to start in the second or third week of June 2010 but the mission was delayed for two reasons. The first delay was operational. ISAF tasked the 1st Brigade, 101st Airborne Division, to support the Afghan defense of Barg-i-Matal in Nuristan. This mission temporarily limited the availability of resources, particularly aviation support. The mission approval process in Afghanistan provided a second reason for delay. An operation of this scope with its need for external aviation support required division (RC-East) approval. This approval took some time and it was not until 23 June that RC-East endorsed the operation. Accordingly, STRONG EAGLE was scheduled to begin on 27 June.[19]

The Plan

STRONG EAGLE was designed to initially secure the village of Sangam, which was in a no-man's land between the opposing sides and then to seize the village of Daridam. Daridam was considered an insurgent stronghold. These objectives were considered modest, reflecting the freshness of the unit in the country and the resistance expected. A follow-on operation (STRONG EAGLE II), tentatively scheduled for July, would clear Chinar, believed to be Rahman's headquarters. In planning, the operation was highly choreographed in order to place supporting elements into overwatch positions at the start of the operation, followed by a main advance down the valley road to Daridam. Vowell did not possess enough strength to provide a blocking force to cover any enemy retreat towards Pakistan. To make up for this deficiency, the TF *No Slack* commander coordinated with Pakistani authorities several hours before the start of the operation. The Pakistanis readily promised cooperation. Rahman was considered a state enemy. The Pakistani border guards prepared to interdict any insurgent movement across the border. Vowell also planned to use air strikes to attack any large bodies of insurgents seen retreating from the battle.[20]

The key element in the operation was a task-organized Headquarters and Headquarters Company (HHC), 2-327th Infantry, commanded by Captain Steven Weber. Weber's command was not a tactical unit by organization, but upon deployment to Afghanistan, HHC had been reinforced with three infantry platoons from other companies, while Weber retained part of his organic heavy weapons and scout platoons. For STRONG EAGLE, Vowell designated Weber as the ground force commander. Weber would directly control about 100 American Soldiers and an equivalent number of

Afghan National Army (ANA) and Afghan Border Police (ABP) forces. His force was codenamed *Team Wolverine*.[21]

For its Afghan deployment, the usually light 2-327th Infantry received a fleet of armored trucks called, generically, Mine Resistant Ambush Protected (MRAP) vehicles. The MRAPs came in a variety of models. The most common type in Weber's force was the Mine Resistant Ambush Protected-All Terrain Vehicle (M-ATV), a state-of-the-art, lighter, more mobile MRAP designed to ultimately replace the up-armored M1114 HMMWV in counterinsurgency operations. The M-ATVs were fresh off the assembly line, the first models having been produced less than six months before. These vehicles provided Weber with a truck that could take RPG and IED hits with high crew survivability and provide heavy firepower through turret-mounted weapons systems. The M-ATVs were equipped with a combination of M2 .50 caliber machine guns, Mk19 automatic grenade launchers, and Common Remotely Operated Weapon Station (CROWS) systems equipped with .50 caliber machine guns. The CROWS system allowed gunners to remotely fire a variety of vehicle-mounted weapons from within a vehicle. However, the newness of the system often proved a hindrance. The CROWS had just been fielded and the 2-327th infantrymen operating them lacked detailed training on the system. Because of an interlinked system design, minor operator errors often shut the system down and when the CROWS went down, so did the whole vehicle. Still, the M-ATVs would be key to a successful advance down the valley road to Daridam.[22]

One unit attached to Weber's command that was equipped with specialized MRAPs was the Route Clearance Package (RCP). The RCP was a hybrid engineer-infantry element containing road clearing equipment including the Husky and the Buffalo. The Husky was a metal detecting and marking vehicle capable of finding and marking metal bombs and mines. The Buffalo was a MRAP designed specifically to clear passage lanes on roads containing mines and boobytraps. Once Weber's dismounted infantrymen secured the road and cleared obvious tripwires, boobytraps, and close-in ambush sites, the RCP was designated to lead and clear the way until Daridam was reached. There Afghan troops would take over in an area where civilians were expected to be found.[23]

Because Coalition policy was that Afghan Soldiers would be the first to interface with the civilian population, Afghan elements played a key role. Platoon-sized elements of ANA and ABP were attached to Weber's force. The Afghans had the mission of leading the advance into civilian population centers. Additional Afghan forces, including a platoon of commandos from Asadabad known as the "Tigers" and a US-sponsored special counterterrorism commando force called "Omega" also supported the operation. Finally, an elite company-sized Special Forces trained Afghan force, codenamed "Zombie" and based out of Jalalabad, was available to support the operation by acting as a clearing force in the later stages of the operation. The Afghan troops were generally mounted in HiLuxes, small pick-up trucks which mounted heavy machine guns.[24]

Vowell wanted surprise for the operation, so that the insurgents would not flee ahead of time and wait for the Americans to leave. Ultimately, he not only wanted to seize Daridam but he also wanted to defeat and disrupt the insurgent network in the valley. Vowell needed to have the enemy stand and fight so that they could be destroyed. This was the main reason he poised his main effort to advance straight down the road: to force Rahman to fight. Since experience from previous tours had taught him that telling the Afghan forces too early about an operation often resulted in the enemy learning about it in advance, Vowell notified the supporting Afghan elements as late as possible. Accordingly, this need for operational security slightly disrupted preparations for STRONG EAGLE. The exact composition of the Afghan forces in support was not clear until the start of the mission. Marshalling the Afghans slightly delayed the beginning of the movement. Additionally, since the Afghans were not briefed thoroughly on the mission, once the action started, they were unable to respond to battlefield conditions clearly. Nevertheless, after STRONG EAGLE Weber maintained that "[operational security] was more important than Partnership... If we had to do it again, we'd still do it the same way."[25]

Lieutenant Colonel Vowell and Colonel Poppas provided for an array of reserves and support elements for the operation. Task Force *No Slack* retained three reserve forces: two US Army platoons and an ANA platoon ready to be airlifted into the battle on short notice. In terms of Army aviation, a scout weapons team (SWT) of two OH-58 Kiowa helicopters, and an attack weapons team (AWT) of two AH-64 Apache helicopters were scheduled to be overhead from dawn to noon. The Air Force would provide on station F15 Eagle close air support during the same period. The Air Force also provided various platforms to provide aerial intelligence, surveillance and reconnaissance (ISR) and electronic warfare (EW) support to the operation.

Operation STRONG EAGLE had five phases (see Figure 4):[26]

1. Team WOLVERINE (Weber) and the ANA/ABP elements would move from Camp Joyce and secure the district center on the outskirts of Marawara. The district center was the marshalling point for the collection

of vehicles moving into the Ghakhi Valley. At the same time, the follow-on force of Tigers and the RCP would consolidate at a tactical assembly area (TAA) located at the junction between the north- south eastside Kunar River road (Alternate Supply Route [ASR] Beaverton) and the Ghakhi Valley road.[27]

2. A platoon each from A and B Companies would be flown by helicopter under the cover of darkness to occupy dominating positions to the north and south of the valley. The A Company force, codenamed *Team Gator* and coming from FOB Monti located near Asmar about 18 miles north of Asadabad on the Kunar River, would occupy LZ Owl (later OP Shaw). The B Company platoon, designated *Team Bayonet* and reinforced with the company tactical command post, was to come in from the southern Kunar Valley, and occupy LZ Hen (later OP Thomas). These positions were between 1000 and 1500 meters from the valley floor. An important part of the plan was that these forces had sufficient firepower with them. The platoons, accompanied by ANA troops, would carry heavy weapons with them —M2 .50 caliber machine guns and MK19 grenade launchers. Vowell hoped to use the two OPs to deny the enemy the ability to emplace heavy machine guns to fire down onto the troops advancing in the valley.[28]

3. A platoon from B Company, 1st Battalion, 327th Infantry, a company attached to Task Force *No Slack* for the duration of the deployment, along with the company tactical headquarters (*Team Bushwacker*), would conduct a dismounted tactical movement from the TAA to secure OP Kahler, a position which overwatched the initial movement of Weber's force along the axis of the valley road from a mountaintop 500 meters to the south. The Kahler position dominated the area around the Marawara District Center, which was the line of departure for Weber's main force.[29]

4. The battalion Scout Platoon, led by Captain Kevin Mott, and reinforced with elements of the Omega force, would be placed by helicopter atop a high hill (Landing Zone [LZ]Hawk/ Attack by Fire Position 1 [ABF 1]) overlooking Daridam. ABF 1 was approximately 700 meters above the valley floor, 500 meters southwest of Daridam. Mott's force was placed to block the insurgent route of retreat from Daridam to the south. LZ Hawk dominated these southerly routes out of the village.[30]

5. Weber's main force would advance down the valley road to Daridam, using a combination of mounted and dismounted movement. Before approaching each populated area (Sangam and Daridam), Weber would place his force into overwatch and allow Afghan forces to lead. The operation's final objective was the clearance of Daridam.[31]

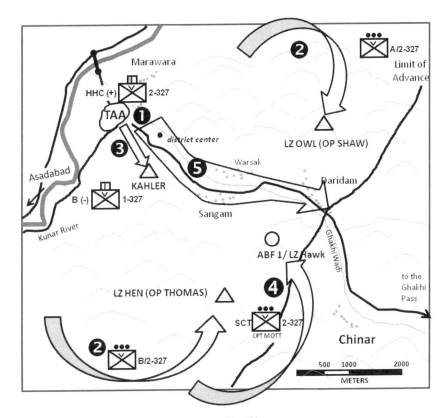

Figure 4. The Plan.

Team Wolverine, under Weber, consisted of two infantry platoons, each equipped with four M-ATVs and augmented with Afghan forces, and his headquarters group. The platoons contained about 40 Soldiers each. Each platoon was divided into a mounted and a dismounted element. Roughly 15 men from each platoon served in the mounted element as drivers and gunners. The headquarters group included several M-ATVs and a 60-mm mortar. The company fire support officer, First Lieutenant Spencer Propst, would coordinate the fires of the Kiowas and Apaches overhead, along with the mortar accompanying Weber and the 155-mm howitzers supporting the operation from Camp Wright in Asadabad.[32]

Weber's two infantry platoons would lead the way. First Lieutenant Douglas Jones's platoon, attached to HHC from B Company, would advance on the left (north) of Weber's axis of advance. Moving dismounted along the northern edge of the wadi north of the road, Jones's force, reinforced with Afghan Border policemen, would clear the hamlet of Warsak northeast of Sangam. He would also cover the left of the force advancing down the road to Daridam. Jones's M-ATVs remained on the road attached temporarily to Weber's headquarters during his dismounted advance.[33]

The other platoon, attached from A Company and led by First Lieutenant Stephen Tangen, with attached ANA elements, was Weber's main effort. Tangen's platoon would move directly down the valley road, advancing dismounted along the southern wall of the valley, to the right of the road. Tangen would be overwatched by his own M-ATVs and those of the group under Weber's direct control. This vehicular force would advance down the road led by the RCP element. The two platoons would move in a bounding manner, with the vehicles supporting Tangen's dismounted element, while Jones would divide his force into two dismounted elements, which would support each other's movements. Tangen's dismounted force would move up to 200 meters in front of the vehicles. After subtracting his mounted element, Tangen had about 25 Americans and 60 ANA Soldiers in his dismounted group. Jones initially had a similar-sized force including ABP attachments. However, after the beginning of the firefight, the ABP troops left and joined their commander who was with Weber on the valley floor, leaving Jones with only the American element. Jones's mission was to cover Tangen's left flank and remove any close-in insurgent threats on the northern edge of the valley, Tangen had the job of clearing the southern part of the valley, including the road, and support the Afghans while they cleared any populated areas, primarily Sangam and Daridam.[34]

The operation was projected to run from about midnight on 27 June to about midday, approximately 12 hours. Vowell wanted to use the cover of darkness to execute his aviation-based maneuvers and so timed the operation so that it would conclude before the midday heat had set in. The attack itself was to begin at dawn. Vowell later commented, "We tried to respect the population there and we did not want to create any problems, so we were going to attack at first light."[35] A daylight move would find the locals already up and at work, and would lessen the possibility of inadvertent civilian casualties.

The Execution

As in most military operations, things did not go according to plan almost from the start. As previously mentioned, there were delays at the beginning of the operation due to the need to marshal the ANA forces participating in the operation. After assembling his units at Camp Joyce, Weber led his convoy north up ASR Beaverton to Marawara village and the district center at about 0230. Enemy ICOM radio intercepts had begun

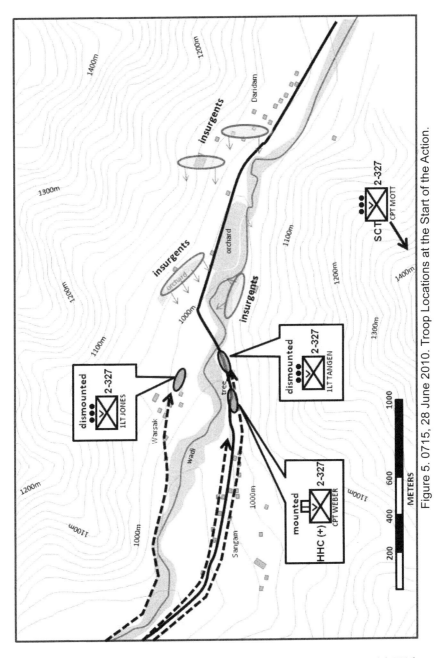

Figure 5. 0715, 28 June 2010. Troop Locations at the Start of the Action.

chattering with news of the anticipated attack as early as 0100. With Weber in place at his start position, Captain Thomas Whitfield's platoon-sized B/ 1-327th Infantry contingent, Team *Bushmaster*, could begin moving to

Photograph taken on 9 August 2010 by Dianna Cahn (Stars and Stripes)

Figure 6. Looking east down the Ghakhi Valley Road at the tree that marked the start of the Ghakhi Valley battle on 28 June 2010. The orchard which was the main insurgent defensive position can be seen off in the distance just under the tree branches in line with the road.

OP Kahler. Whitfield's dismounted force started its climb at 0300. Under normal circumstances the trek to the peak summit that was Kahler took 45-minutes. Weber's advance was contingent on Whitfield being in place to overwatch his initial movement.[36]

The helicopter borne over-watching forces were established at LZs Owl and Hen at the same time *Bushmaster* started its movement. With these forces in place, and *Bushmaster*'s movement taking longer than expected, Vowell felt safe to let Weber begin his movement even without Whitfield's occupation of OP Kahler. After a delay waiting for the ANA elements to come forward, Weber began his movement around 0615. The initial advance through the village of Sangam was rapid and virtually uneventful. The Afghans quickly cleared each building as there was no opposition and no civilians were present. On the left, Jones's group, moving through rough, hilly terrain, cleared Warsak without any contact. As at Sangam, there were no civilians in Warsak. The local population had received enough advance warning to flee to the mountains before the operation started. This warning was undoubtedly spread by word-of-mouth from those villagers in the west who saw the assembly of the Coalition forces near the district center.[37]

The quiet was deceptive. While still in Sangam, Tangen heard a report that ICOM intercepts indicated that a force of at least 40 insurgents awaited him in Daridam. Weber and Tangen felt that they had sufficient firepower and supporting elements to deal with this force if it materialized. In addition, the Kiowas of the SWT, which were flying overhead, had already begun to take sporadic AK-47 rifle fire from positions near Daridam. Followed by Weber and the RCP, Tangen's group cleared Sangam by 0700. Lieutenant Colonel Vowell planned to observe the initial movements via a command and control helicopter over the battlefield. Once the initial movements were completed, he then intended to dismount and join his forward elements on the ground. However, the mountainous terrain hindered radio communications and Vowell was forced to stay aloft to act as a relay station between Weber and the battalion tactical operations center (TOC) at Camp Joyce.[38] The action started in earnest at 0715 when Tangen's dismounted group passed a large tree hanging across the road beyond Sangam (Figure 6). Weber later speculated that the tree marked the trigger line for the enemy forces. The tree was at a point in the road just before it crossed the wadi. Here there was no shoulder to the south, as the slope began right at the road. Tangen advanced on foot with Staff Sergeant Eric Shaw's squad leading, followed by Tangen and the squad of Staff Sergeant Robert Livingston. The firefight began as soon as Tangen's men passed the tree. Two RPG rounds exploded nearby, followed by PKM machine gun fire. Both fusillades came from Tangen's front in the direction of Daridam. Despite the positioning of forces on the flanks, the insurgents still retained excellent positions on the high ground around Daridam and fiercely resisted the American attack. Tangen's force was trapped with nowhere to go, as there was no cover on the side of the road. The ANA troops, who were in the lead, took immediate casualties. Squad Leader Shaw ran forward to assist the ANA commander in responding to the contact. Meanwhile the initial volley had also wounded Private First Class Stephen Palu. Livingston and Corporal Joshua Frappier bandaged Palu's wounds.[39]

Jones's group to the north also came under fire, but the fire was of a lesser intensity than that facing Tangen. Neither Jones nor Weber could easily discern the source of the enemy fire. This lack of positive identification of enemy firing positions hindered the use of supporting and direct suppressive fires. Nevertheless, carrying the wounded ANAs and Palu with them, Tangen's element was able to bound back to the vehicles, which had advanced as far as the tree. The RCP element pushed forward to meet Tangen's men halfway. The vehicles covered the retreat of the dismounted group with suppressive fires. Tangen's men then used the trucks as cover while directing the fires of the mounted weapons at suspected enemy positions.[40]

At this time, before he could reach safety behind one of the M-ATVs, Staff Sergeant Shaw received a fatal wound to the head. In the meantime, Captain Weber had gone back to Sangam to get a HiLux truck from the Tiger element to use for MEDEVAC. The bed of the truck and its heavy machine gun made it a good choice for this role. The HiLux arrived to evacuate the casualties. The tally at this point was one American killed in action (KIA) and one wounded in action (WIA), four ANA WIAs and one ANA KIA. Concurrent with the casualty evacuation, the remaining ANA troops with Tangen and Weber withdrew and retreated with their wounded to Marawara. The Afghan Border Police element remained with Weber. The first phase of what would become an all-day firefight was over.[41]

Weber, now located back in Sangam, worked to reorganize his force in preparation for continuing the advance as the day started to warm up. There was a traffic jam of vehicles along the valley road stretching back from Sangam to Marawara. The HHC commander had to get the most important vehicles to the front of the column. These included the specialized vehicles of the RCP, additional M-ATVs with their armored protection, and the Tiger force's HiLuxes with their heavy DShK machine guns. He tried to get the bulk of the RCP forward to support Tangen, with mixed results, as the RCP had problems on the rocky road. Then as the unit advanced up the road to the front, an RPG hit one vehicle. At this point the RCP platoon leader felt his unit was combat ineffective and decided to withdraw to Marawara. Weber and his subordinates had to use all their powers of persuasion to keep the RCP in the fight. [42]

To reduce the enemy pressure on the lead unit, Propst, the Fire Support Officer, had the 155s at Asadabad drop smoke rounds between Tangen's group at the tree and Daridam. At the same time, Apaches from the Aerial Weapons Team began conducting gun runs against insurgents entrenched in a pomegranate orchard located north of the road. The Apache runs continued for over two hours through mid-morning. The heat began affecting the troops as much as the enemy fire. As the day went on, most of the company leadership, including Weber and Propst, needed intravenous (IV) infusions to quell the effects of dehydration.[43]

While Weber paused, the forces on the ridgelines to the north and south also found themselves in firefights with insurgent forces attempting to

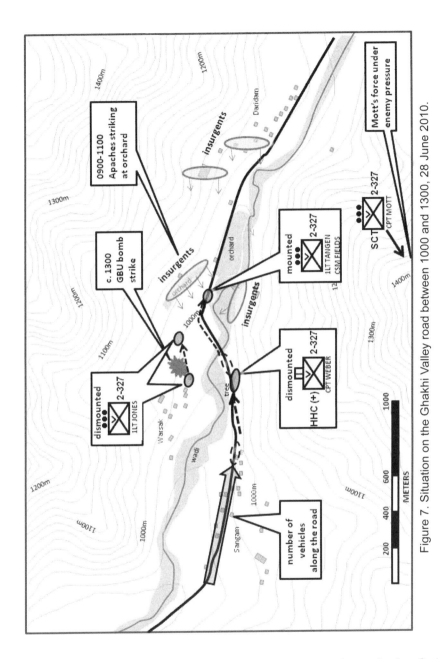

Figure 7. Situation on the Ghakhi Valley road between 1000 and 1300, 28 June 2010.

outflank the units on the valley floor. The enemy seemed surprised to find Americans high on the mountains. To the north, the A Company platoon at LZ Owl ambushed an enemy element trying to move to the location of the overwatch position, unaware that the Americans were there. After this success, the enemy kept the position under heavy fire. Similarly, the B

111

Company position, (LZ Hen) to the south of Sangam was heavily engaged. The Team *Bayonet* commander later felt that the position, although on the highest ground in the area, still had some weaknesses that allowed gave the insurgents some advantage. "Positions on highest points [often] result in lots of dead space." [44] The amount of firepower the insurgents were able to place on LZ Hen resulted in the death of Sergeant David Thomas by gunshot wounds. Eventually, the Aerial Reaction Force (ARF), a platoon under the C (Cougar) Company commander, reinforced, then relieved the Bayonet forces at LZ Hen. [45]

To the northeast of B Company's position, the Scout/ Omega force at its attack by fire position (LZ Hawk) overlooking Daridam from 1500 meters was in trouble. That group, under Captain Mott, was surrounded and cut off from support even from the relatively nearby B Company position. In the midst of heavy enemy fire Mott received a grazing gunshot wound to the head and then fell 100 to 200 meters down steep ground. Miraculously he survived. Mott's group had had a tough go so far in STRONG EAGLE. Initially the helicopter had dropped the group off on the wrong ridgeline. In the darkness the scouts trudged as near to the correct position as they could reach under the cover of darkness. Mott divided his force in half, placing each on mutually-supporting hilltop positions. He led one of the elements, his platoon sergeant, Sergeant First Class John Howerton, the other. Eventually, commandos from the Omega element joined Mott and took up positions in the middle between the two scout sections, providing a unified defensive position. When the action started in the valley, Mott's group began receiving effective fire from every direction. [46]

After Mott fell, his men soon realized he was still alive. In fact the platoon leader managed to drag to cover under a tree. For anyone to try and reach him at the time would have been suicidal given the intensity of the enemy fire and the angle of the slope. Mott could not be evacuated until later. Meanwhile Howerton assumed leadership of the force overwatching Daridam. [47] It was now around 1000. From listening to the radio traffic, Weber realized the forces on the high ground were being pressured by the enemy. He felt it was imperative to push forward to decrease this pressure.

Given the open nature of the ground in front of Tangen, where the road crossed the wadi, Weber decided a mounted advance was proper. Four of Weber's M-ATVs had snaked their way through the column and reached him at Sangam. These he sent forward to Tangen's position near the tree. Weber himself followed dismounted, joining Livingston's squad near the big tree and positioning himself inside a nearby house from which he could see and control the movements of both Tangen's and Jones's groups. [48]

Tangen began moving forward with the bulk of his dismounted force now mounted in four vehicles. The group crossed the wadi and took the big bend in the road that turned east towards the orchard and Daridam beyond. Meanwhile, Jones began moving towards the open area beyond the end of Warsak. Afghan HiLux trucks supported the advance from the vicinity of the tree with their heavy machine guns. Smoke fired from the 155s at Camp Wright covered this movement. When Jones reached the outskirts of Warsak, he was confronted with a large open area. He paused in the shelter of the last few buildings, expecting that the actions of Tangen's group and supporting ground and aerial fires would cause the enemy fire to slacken.[49]

Weber began advancing dismounted toward the big bend in the road with his command group and Livingston's squad. He realized that his group was now taking sporadic fire from a group of insurgents positioned to the south along the aqueduct at the southern edge of the wadi. Suddenly Tangen's lead M-ATV was hit in the engine by one or more RPGs fired from the orchard to the left front of the road and a compound located just beyond the trees. The armored plating on the vehicle allowed the crew to survive with no injuries. However, the vehicle was disabled along with its CROWS weapons station.[50]

It was now about 1100. At this point Command Sergeant Major Chris Fields, the battalion command sergeant major, approached Weber's position with a small MRAP convoy. Weber was showing the effects of little sleep and the extreme heat. Nevertheless he felt he had to push forward to keep the advance moving. Fields volunteered to take on that mission and advised Weber to establish a stationary command post and recover from the effects of the heat. Weber accepted the advice and Fields moved forward to join Tangen. Time passed as Tangen and Fields tried to get the M-ATV off the side of the narrow road or set it up to be towed. In the meantime, the firefight continued with several of Tangen's NCOs killing with hand grenades some insurgents who had crept close to their position.[51]

As Tangen and Fields sought to clear the narrow road, Jones's men remained at the edge of Warsak. At about 1300 it seemed that the effects of Apache gun runs and other supporting fires would allow Jones to move forward and assault the orchard to his front. Apaches were also firing Hellfire missiles danger close to Jones's men, into the orchard and the compound behind it. Jones began moving his element forward. The platoon was divided into two groups. Jones personally led one force, which he advanced across some open ground over a spur to a house laying beyond the edge of Warsak. The other half of his platoon covered this move from a house at the edge of the hamlet. Before Jones could bring the trailing element up to his location, Weber called him and warned him that a danger close air strike was about to hit in the nearby orchard.[52]

This air strike surprised Weber as much as it did Jones. For the STRONG EAGLE Operation, Task Force *No Slack* had numerous nonstandard attached units. Some of these forces were elite Afghan units supported by or partially staffed with American advisers. Although he had overall tactical command of the forces in the operation, Vowell had designated Weber as the ground-force commander. As such, Weber, with his unique situational awareness, controlled the supporting fires. He knew where his men were and would not have called in fire so close to his own forces. It is possible that one of the adviser teams with the Afghan units in the valley made the call without precise knowledge of where US troops were at the time.

Both Jones and Tangen were accustomed to the danger close Hellfire missile strikes impacting on nearby enemy positions. But the GBU-12, carried by supporting US Air Force F-15 jets, with its 500 pounds of explosive, was a much larger ordnance package. The GBU had a bursting radius of about 170 meters. This was fairly close to both Jones's and Tangen's groups. Weber warned both officers on the radio and the lieutenants quickly barked orders to their subordinates to take cover. Jones was initially concerned about Tangen's men, who were out on the road with little cover to protect themselves from the shrapnel effects of the pending airstrike. Tangen's men were also closer to the orchard target. Both groups took cover. Tangen's men went inside vehicles and into a small ditch along the south side of the road. The 15 men in Jones's rear element huddled in the ruined house at the edge of Warsak. Jones and his contingent used the protection of the nearby building they had just reached.[53]

In spectacular fashion, the GBU exploded and rocked the terraces north of the valley road in front of the orchard strongpoint. The bomb landed between Jones's two groups, closer to the rear element. A post battle investigation revealed that the grid coordinates called for the air strike were only six meters east of the rear detachment's position. Jones himself was 70 meters east of the impact. He feared he had lost half his platoon in one stroke. Fortunately, the modest protection provided by the ruins was enough to protect the men from the full effects of the bomb. Five Soldiers were wounded, three requiring immediate evacuation. The other two would be eventually evacuated, but initially assisted in bringing the others down to Weber's position for medical evacuation.[54]

The air strike effectively halted American operations for the rest of the day. Both Jones and Weber became consumed with the medical evacuation (MEDEVAC) of Jones' wounded men. Weber used a helicopter already scheduled to drop a resupply of water to the forward positions for the MEDEVAC mission. The flat ground next to the road near the wadi provided a convenient landing zone. Jones first evacuated the three badly wounded men. They were on their way to medical facilities in Asadabad within 25 minutes. Afterwards, the helicopter was recalled to evacuate the other two men. During the second MEDEVAC, enemy fire began to increase. Soon insurgent machine gun fire knocked out a second M-ATV in Tangen's group, striking the radiator and causing the engine to overheat. Weber made the decision to abandon the damaged vehicles. The Soldiers quickly removed sensitive items from the damaged vehicles and pushed them to the side of the road. Tangen's men then withdrew to the wadi where they joined with Weber's group and set up a defensive position. From this location they could overwatch and cover the abandoned vehicles with fire. No one attempted to approach them. Following the disabling of the second M-ATV, insurgent fire again slackened. This decrease showed the effects of the cumulative hours of Apache and field artillery strikes. After the completion of the MEDEVAC missions, the increasing effects of the fatigue of combat and the heat forced Weber to halt his men where they were. It was at this time that Weber himself had to receive intravenous infusions to keep from becoming a heat casualty. HHC had a total of one KIA and six WIAs in the day's action so far. Weber told his commander, Lieutenant Colonel Vowell, that he estimated that his men would need three or four hours to get ready for renewed operations.[55]

Battalion commander Vowell had been aloft all day. He realized that Weber's force was exhausted and needed rest and refitting. If he expected to capture Daridam, he needed to come up with a new plan. He landed at the district center outside of Marawara and, after ordering Weber to hold his current positions, he contacted the brigade commander, Colonel Poppas. Poppas had been following the battle from his headquarters in Jalalabad and had already come up with a solution. The brigade commander committed the Jalalabad-based Zombie force to the battle. The Zombies, a group of about 80 elite Afghan commandos advised by a US Army Special Forces Operational Detachment A (ODA) would fly up to Marawara on three Chinook helicopters and assume the mission of the main effort from Weber.[56]

The new plan (Figure 8) directed the Zombie element to attack after midnight, supported by Weber's men. The attack would begin after an Air

Force AC-130 Spectre gunship arrived to provide fire support. Vowell met the Zombies when they landed and provided them with a brief situation update. The Zombies moved up the road to Weber's position in the wadi, arriving there before midnight. Initially the Afghans thought the recumbent Soldiers were casualties and were pleasantly surprised to see that they were just resting. Weber still believed that he needed a little more time before his men would be ready to go. Unlike the ANA that he had dealt with earlier in the day, Weber found the Zombies eager to go forward. They did not want to wait for the Americans to recover fully.[57]

Accordingly, Weber drew up a hasty plan for the attack on Daridam. Instead of advancing along the road, the commandos would move down the wadi into the town, which they would then clear. Jones and Tangen would support the advance by fire from their current positions then follow up by clearing the orchard and road. The distance to Daridam from Weber's position was about 600 meters. There was some confusion as Weber thought the Zombies intended to clear the whole village, but from his later actions, it was clear that the Zombie commander only intended to gain a foothold in the village.[58]

The commando force began moving easterly down the wadi at about 0030 on 29 June. Almost immediately enemy forces in the orchard and in several compounds north of the road opened fire on the attackers. Vowell had placed the Spectre gunship on alert. He felt free to do this because he had numerous indicators that there were no civilians in Daridam, just as there had been none in Sangam or Warsak. After spotting targets, the aircraft began firing on the insurgent positions, providing Weber's men with a spectacular light show. [59]

The Spectre fire was so obviously effective that Weber decided to send his Soldiers forward immediately. He sent a mounted detachment under the company first sergeant down the road while simultaneously setting Jones and Tangen in motion dismounted. Tangen had orders to clear the orchard. Jones would advance on Tangen's left and clear the compound beyond the orchard that had been an insurgent stronghold. The operation was successful. By dawn the Americans joined the Zombies at their foothold on the western edge of Daridam. At this time it became readily clear that the Zombie commander did not intend to clear the rest of the village.[60]

Vowell arrived on the scene with the ANA and ABP forces that had been part of the original operation. At this point it was readily clear that the enemy had evacuated Daridam. The Afghan forces were now eager to collect war booty. In this, competition between the ANA and the ABP

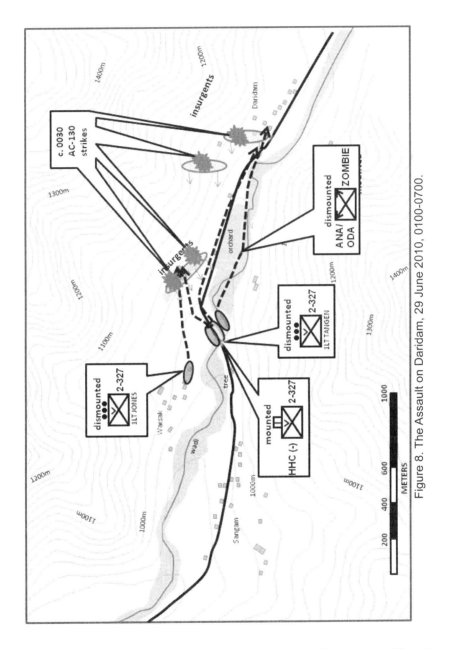

Figure 8. The Assault on Daridam, 29 June 2010, 0100-0700.

emerged. The two Afghan elements immediately began searching the village. Tangen and Jones followed in support. Although there was no actual fighting in Daridam, several remaining insurgents fled the village only to be engaged by the Kiowas of the SWT.[61]

TF *No Slack* had faced an adversary in what was the worst case scenario devised by the battalion staff. Post-battle estimates were that at least 250 insurgents faced the Coalition force in the Ghakhi Valley. Rahman had committed virtually his whole force. Although only a few bodies were found in the orchard, estimated enemy KIAs, based on aerial observation and blood trails, were at least 125.Vowell considered that his S2 (intelligence) staff section had done a great job anticipating Rahman's reactions. The tough fights faced by the units on the hilltops north and south of the valley demonstrated this. Despite the accurate projections, the fight was still harder than expected. The rugged terrain compartmentalized the battle, transforming it, at least initially, into a series of smaller, separate actions, each opposed by a determined enemy force. The fierce resistance made the battle last longer than anticipated and the 115 degree temperature exhausted the attackers, requiring reinforcement and rest in the midst of combat.[62]

The Aftermath and Strong Eagle II

Marawara remained a dangerous place for non-combatants. On 28 June, while Weber's forces were fighting it out with the insurgents on the outskirts of Daridam, the Marawara District intelligence chief was killed when his vehicle struck an IED. The 2-327th battalion TOC displaced directly into Daridam on 29 June. The Americans stayed in Daridam for four days, before handing over control of the village to the Afghan police. American troops continued to man LZ Hen, renamed OP Thomas, along with ANA Soldiers, in preparation for the continuation of the advance to Chinar. The observation post was fortified. The valley below however, remained unoccupied by the enemy. [63]

In the immediate aftermath of STRONG EAGLE, tribal leaders from the district held a meeting in Asadabad where they pledged support to the government, while requesting a permanent security presence throughout the valley. The civilians gradually returned to the eastern valley and the Kunar PRT started or renewed reconstruction and development projects. The chain of command quickly recognized the valor and fighting abilities of the American Soldiers in the battle. Six Silver Stars and a Bronze Star Medal were awarded to members of 2-327th Infantry personally by Secretary of Defense Robert Gates in December 2010. US Afghan theater commander General David Petraeus awarded an additional Silver Star and two Bronze Star medals for the operation on 6 January 2011.[64]

In the days immediately after STRONG EAGLE's completion, insurgents attacked Coalition outposts at Asmar (FOB Monti), Camp Joyce and several other places in Kunar Province. The attack on FOB Monti was particularly unusual as that area was usually quiet. Vowell later reflected that these attacks were an effort by Rahman to draw forces away from Marawara in the aftermath of STRONG EAGLE. With his main force bloodied in the Ghakhi Valley, these attacks were seemingly conducted by local forces, not those under Rahman's direct command. Nevertheless the attacks suggested that the enemy commander had the ability to control forces throughout the Kunar Valley, even in the midst of a battle at his front door.[65]

After a respite of several weeks, in order to keep the pressure on Rahman, Task Force *No Slack* executed Operation STRONG EAGLE II, a limited objective attack designed to capture the village of Chinar on 19 and 20 July 2010. Chinar was east of Daridam and was considered to be Rahman's stronghold, even during the June battle. Company B, 1-327th Infantry, attached to Task Force *No Slack* and based at Asadabad, led the new advance. The second offensive was patterned on the first, with synchronized overwatch forces and readily available reserves. One lesson Vowell had learned from the original operation was the usefulness of a night attack. With the civilian population showing an ability to flee before any combat action, he felt that the fear of civilian casualties in a night move was minimal. Accordingly, the STRONG EAGLE II assault began in the middle of the night, not at dawn. Vowell advanced a force of approximately 450 Americans and Afghans against the village. After the fierce resistance encountered previously, the Americans expected more of the same. However, the enemy either did not want to take similar losses, or no longer had the available strength to oppose the Americans. Rahman's force withdrew from Chinar before the Americans arrived. The attackers encountered only minor resistance. [66]

Redevelopment and reconstruction in Marawara was uneven. Immediately following STRONG EAGLE II, the village elders participated in a district *shura* held in Daridam in which they promised to support Coalition redevelopment efforts in the district. But this promise depended on continued security in the valley. In this regard, some of the weaknesses in the Afghan forces shown during the battle persisted afterwards. On 28 July US forces handed over security in the valley to the Afghan National Police. However, the Afghans were fearful of staying in the valley without American forces. Accordingly, US forces occasionally returned to the area. In early August US and Afghan forces conducted raids aimed at capturing or killing Rahman. The elusive Rahman escaped, although up to 30 of his fighters were killed in the actions.[67]

With Marawara relatively quiet after the beating Rahman's force had taken in STRONG EAGLE, Vowell needed to use his forces elsewhere. In the restive Ganjgal Valley just outside Camp Joyce, the 2-327th IN conducted a series of operations (EAGLE CLAW I and II) in the fall and winter of 2010. US forces returned to Marawara in force in October 2010. In the aftermath of the unsuccessful rescue operation for kidnapped British aid worker Linda Norgrove, which took place to the west in the Pech region, US Special Forces fought the kidnappers in Marawara as they attempted to retreat into sanctuary areas in Pakistan.[68]

Conclusions

Vowell's battalion had just arrived in Afghanistan when STRONG EAGLE was executed. The success of the operation shows that in the ninth year of the War in Afghanistan, Army leadership and Soldiers were so experienced and well trained that while 2-327th Infantry was new to the country, most of its leaders and many of its Soldiers were familiar with the country and the enemy. Vowell had served previously elsewhere in Afghanistan and had talked to officers and read reports from his two predecessor units in the Kunar Valley. Based on this pre-deployment preparation, he was able to plan and execute a large operation in a vital district almost as soon as he assumed responsibility for eastern Kunar. The enemy no longer could depend on a respite when new units arrived in theater were still unfamiliar with the area of operations.[69]

Despite the heavy fighting, US casualties were relatively light. This was a tribute to both planning and technology. Lieutenant Colonel Vowell's seizure of the high ground with significant forces prevented the enemy from pounding Weber's men from positions with clear fields of fire. The use of night air assaults into the mountains surprised the enemy who were ambushed in several places while trying to move against Weber's force. Rahman's supporting elements were forced to fight the troops on the high ground instead of those in the valley. These fights were as fierce as those below and played an equal part in the success and destruction of the enemy force that resulted in the later success of STRONG EAGLE II.

As mentioned previously, the positioning of the observation posts in the high ground proved effective. In fact, American forces were located at LZs Owl and Hen surprised the insurgents. In the case of Owl, the enemy walked into an ambush. However, the insurgents soon recovered from their surprise and fought hard to eliminate the positions. As also previously discussed, the foresight in bringing heavy weapons to the hilltop positions proved to be decisive as the defenders were able to gain and maintain fire superiority over the attacking enemy. The enemy forces tied down on the heights were unable to participate in the battle on the valley floor.[70]

In terms of general combat verities, the battle outside Daridam displayed once again the importance of fresh reserves and the general exhaustion troops can suffer in extended firefights, particularly in a hot environment. Despite an accurate assessment of enemy capabilities and intentions, and, despite the execution of a plan designed to account for these capabilities and intentions, the fight was still tough. Success still required the commitment of fresh reserves. Vowell showed flexibility during the action in modifying his plan once it broke down under enemy resistance and weather conditions. The TF *No Slack* commander watched the battle without directly interfering until necessary to ensure that the reinforcements were used properly. Both Vowell and Weber realized that exhaustion had overcome the advanced troops and that these men required a period of regrouping and reinforcement to continue the attack.

Technologically, the MRAP concept proved its worth. Clearing the road in the teeth of an entrenched enemy defense that was less than 200 meters away, the advance guard of Weber's force suffered only one KIA despite being in the presence of the enemy for over 12 hours. The ability of the Army logistical system to rapidly field newly produced M-ATVs to frontline units in Afghanistan was remarkable. The newness of the equipment resulted in some familiarity difficulties during the action, which were more than balanced by the ability of the vehicle system to survive enemy fire. The M-ATVs that were disabled were knocked out because of RPG and small arms hits against the engine. Future developments of the vehicle need to account for this and make it easier for troops under fire to recover from engine failures or more easily tow the vehicle off a narrow road.

The use of the MRAPs in STRONG EAGLE is reminiscent of the use of tanks in support of infantry in other conflicts. Despite the stigma of large numbers of Soviet tanks operating ineffectively in Afghanistan, this does bring up the question of whether the deployment of main battle tanks to Afghanistan may have some utility in certain areas and for certain types of operations. Tanks could have surely operated on the Ghakhi Valley road and would have been more survivable than the MRAPs. An RPG would not stop an M1 tank by a hit to an exposed engine. Tanks also would have had tremendous on board direct firepower to use against personnel and point targets. While tanks may have had limited mobility in certain parts of Afghanistan, the Ghakhi valley was not one of those places.

Despite the success of the MRAP, technology is not a panacea for modern combat. Coordination between a mix of forces from various governmental agencies and nations is complex and ripe for confusion and mistakes. The air strike that landed close to Jones's men showed the importance of fires being approved by the commander on the ground that has direct knowledge of the location of the most forward friendly forces. In the last several decades, American forces have developed digital systems that provide virtual complete situational awareness for American forces. However, the terrain in Afghanistan often hinders digital communications. Additionally, most actions are combined operations with Afghan forces that are not equipped with the new technology. STRONG EAGLE also employed several elite Afghan units (Tiger, Omega, and Zombie) which did not operate under the usual chains of command. This characteristic of the force increased the complexity of command and control in spite of technological advances. Vowell himself spent much of the battle in a command and control helicopter coordinating communications. Weber, the designated ground forces commander, found himself in the frontline at various times and, while this gave him visual situational awareness of his dismounted elements, it also forced him to do necessary coordination under duress. Coordination remains difficult at best under combat conditions. Even with an Omega liaison officer by Weber's side, this did not prevent an Omega observer on the OP overlooking Daridam from calling in an uncoordinated air strike. This event was a direct result of the incorporation of forces outside the typical chain of command into the operation. Commanders at all levels need to train, plan and coordinate for such eventualities.

Sometimes the simplest technology is still the most effective. For much of the action, Afghan Tiger forces supported American forward elements by fire with DShK heavy machine guns mounted on the beds of Toyota HiLux pick-up trucks. While the HiLuxes did not have the survivability of the MRAPs, they were able to travel in the most rugged of Afghan terrain and bring to bear large amounts of firepower rapidly. There were numerous HiLuxes on hand as the vehicle was ubiquitous in Afghanistan, not only in the military forces, but also among the civilian populace.

Other types of supporting fires were also key ingredients to the success of STRONG EAGLE. The stationing of SWT and AWT helicopter patrols over the Ghakhi Valley continuously during the operation meant that direct fires were readily available to Weber and his subordinates. A two gun platoon of 155mm field artillery from nearby Asadabad was most useful in providing smoke to screen movements at key times and places. Most critical was the use of the nocturnal AC-130 Spectre gunship. Although employing essentially 1960s era technology, updated with 1990s era targeting equipment, the Spectre was able to accurately place a large volume of fires onto enemy positions under the cover of darkness. The sheer amount of firepower revitalized the American forces and resulted in the rapid clearance of Daridam. So much fire support was available that Weber never used the 60mm mortar he brought with him.[71]

The use of supporting fires was initially tempered by a fear of civilian casualties. Once Vowell and his subordinates discovered the local residents had fled their homes, this support was used based primarily on military considerations, rather than being restricted in its use out of a fear of inadvertently causing civilian casualties. A combination of aerial and ground fires allowed Coalition forces to withstand close-in defensive firepower in the valley and direct attacks on the hilltop positions with a relatively small number of casualties.[72]

STRONG EAGLE showed the enemy at his best and at his worst. According to intelligence estimates, which Vowell considered to be remarkably accurate, Qari Zia Rahman committed all of his forces to the battle outside of Daridam. Resistance was fierce. The insurgents fought well from pre-prepared positions that were able to withstand the pressure of American fire support for several hours. They were able to use the irrigation/aqueduct system in the valley to both remain supplied with water, and to provide concealed fighting positions.[73]

However, Rahman followed a pattern familiar to those who conducted previous forays in the Ghakhi Valley. While defending the valley in front of Daridam from entrenched positions, he also planned to occupy dominating hilltop sites to surround the forces advancing down the road and destroy them with firepower coming from all sides. This very design had worked well against the Soviets in 1985. Vowell was able to successfully counter this typical blueprint with a night insertion onto the high ground that surprised the enemy.

Additionally, while enemy fire was intense, it was not particularly accurate. When the action opened, Tangen's group was caught dismounted along the road in a predetermined kill zone where the terrain provided no cover. Nevertheless, he and his men were able to retreat up to 200 meters to the protective cover of the trailing vehicles while suffering only one KIA and one WIA. The 155mm smoke rounds helped to screen effectively these and later movements. All forward coalition elements spent long periods of time at close to medium direct fire range of enemy fighters and did not lose their ability to maneuver and did not incur heavy casualties.

Perhaps the most difficult aspect of STRONG EAGLE was the achieving of military objectives within or near a civilian population. Fighting battles among a civilian population without incurring unnecessary civilian casualties is difficult. Vowell initially planned his operations to minimize such collateral damage. While initial movements were planned for darkness, the advance was slated for dawn, when civilians could be readily recognized and would be awake and going about their business. A daylight fight also offered a defending enemy certain advantages, especially the ability to see Coalition forces as they approached. Vowell saw this as a tradeoff to save civilian lives. Once the battle started however, it was clear that the local population had enough advance warning to flee the area ahead of time. Both Sangam and Daridam were deserted. The only people in the area were insurgent fighters.[74]

Vowell also showed flexibility in the later stages of STRONG EAGLE and in STRONG EAGLE II once he realized that civilians were not a factor. The final assault and the close air support provided by the Spectre took place at night as did the initial assault in STRONG EAGLE II. In counterinsurgency operations, while the presence of a civilian population puts certain restraints on combat operations, commanders have to be able to recognize when these restraints are not necessary and act primarily based on military considerations.

STRONG EAGLE displayed the ability of American commanders to mass forces for specific combat operations in the midst of a counterinsurgency campaign and to take the fight to places which the enemy felt were safe from Coalition interference. In later operations (STRONG EAGLE II and STRONG EAGLE III), TF *No Slack* shifted the battle farther to the east in Marawara District, almost to Pakistan border as US and Afghan forces eventually fought to clear Rahman's hometown of Barawolo Kalay and the neighboring village of Sarowbay. Both villages are located on the slopes of the Hindu Raj Range, one north-south valley west of Pakistan. The orchard outside Daridam has now become a Coalition agricultural project.[75]

Despite the ongoing efforts by the Coalition, the counterinsurgency campaign in Marawara District is not yet complete. However, the vigorous offensive action by a fresh unit in Operation STRONG EAGLE meant that the next time the enemy stood and fought in the district, it would be at its far eastern edge, against the Pakistani border and in sanctuary areas long felt to be out of the reach of Coalition forces.

Notes

1. Wesley Morgan, "Afghanistan Order of Battle, August 2010," Institute for the Study of War, www.understandingwar.org/files/AfghanistanOrbatAug10.pdf (accessed on 14 April 2011); "US Contributions and OEF/ISAF C^2 in Afghanistan 2001-2010," PowerPoint briefing, 16 November 2010.

2. Morgan.

3. For a detailed discussion of operations out of Asadabad during this period see Master Sergeant Michael D. Coker, interview by Major Doug Davids, Combat Studies Institute, Fort Leavenworth, KS, 9 August 2007, http://cgsc.cdmhost. com/cdm4/item_viewer.php?CISOROOT=/p4013coll13&CISOPTR=665&C ISOBOX=1&REC=19 (accessed on 9 February 2011); and Lieutenant Colonel Anthony Shaffer, *Operation Dark Heart* (New York: St. Martin's Press, 2010).

4. For more information on conventional operations in Kunar Province during this period, see Major Rich Garey, interview by Laurence Lessard, Combat Studies Institute, Fort Leavenworth, KS, 5 November 2007, http://cgsc.cdmhost. com/cdm4/item_viewer.php?CISOROOT=/p4013coll13&CISOPTR=996 (accessed on 9 February 2011). The outpost at Barikowt was closed in 2003. For more information on MOUNTAIN RESOLVE, see SGT Greg Heath, "10th Mtn. Div. Shows its Mettle in Operation Mountain Resolve," *Defend America*, 17 November 2003, http://www.defendamerica.mil/articles/nov2003/a111703e. html (accessed on 10 February 2011) and http://www.defendamerica.mil/articles/ nov2003/a111703f.html (accessed on 10 February 2011).

5. The Marine battalions were 2/8th Marine Regiment (November 2003-May 2004); 3/6th Marine Regiment (April-December 2004); 3/3d Marine Regiment (October 2004-June 2005); 2/3d Marines (June 2005-January 2006); and 1/3d Marine Regiment (January- June 2006). For an account of the operations of the 2/3d Marines see Ed Darack, *Victory Point: Operations Red Wings and Whalers--The Marine Corps' Battle for Freedom in Afghanistan* (New York: Berkley Caliber, 2009).

6. Kent Harris, "2nd Battalion, 503rd Infantry Regiment: New Deployment, New Plan of Action," *Stars and Stripes* (10 December 2010), http://www.stripes. com/news/2nd-battalion-503rd-infantry-regiment-new-deployment-new-plan-of-action-1.128192?localLinksEnabled=false# (accessed on 22 March 2011); Lieutenant Colonel Joel Vowell, interview by Captain Sterling L. DeRamus, 2010; Bing West, *The Wrong War: Grit, Strategy, and the Afghanistan* (New York: Random House, 2011), 52, 91, 127.

7. West, 52; Interdiction of enemy forces as a function of counterinsurgency operations is discussed in Department of the Army, *Field Manual 3-24, Counterinsurgency* (Washington, DC: Department of the Army, 2006), pages 5-19, 5-29, 8-5, and E-5.

8. Ghakhi itself is a village east of the pass in Pakistan.

9. Tony Wickman, 'Building a Road and Bridging a Gap in Kunar Province,

" PRT-Kunar website, 22 October 2009, http://prt-kunar.blogspot.com/2009/10/
building-road-and-bridging-gap-in-kunar.html (accessed on 29 March 2011);
Nicholas Mercurio, "PRT Awards Marawara Bridge Contract," CJTF-101
website, 10 March 2011, http://www.cjtf101.com/en/regional-command-east-
news-mainmenu-401/4244-prt-awards-marawara-bridge-contract.html (accessed
on 29 March 2011); Vowell interview; Brian Boisvert, "The Dirt Boys of ADT,"
PRT-Kunar website, 24 November 2009, http://prt-kunar.blogspot.com/2009/11/
dirt-boys-of-adt.html (accessed on 29 March 2011); Vowell interview; Charles
Brice, "Team Works to Improve Irrigation in Afghanistan,"*American Forces Press
Service,* 30 December 2008, http://terrorism-online.blogspot.com/2008/12/team-
works-to-improve-irrigation-in.html (accessed on 4 April 2011). After STRONG
EAGLE, summer rains in 2010 caused flooding that damaged the bridge. This
forced its reconstruction in 2011.

10.   Russian General Staff , *The Soviet-Afghan War: How a Superpower
Fought and Lost*, trans. and ed. Lester Grau and Michael Gress (Lawrence, KS:
University Press of Kansas, 2002), 195-6; Vladimir Grigoriev, "Facts of War
History: Marawara Company," Art of War, http://artofwar.ru/g/grigorxew_w_a/
text_0080.shmtl (accessed on 23 March 2011). The Soviets called Asadabad by
its old name of Chigha Sarai.

11.   Grigoriev; Lester A. Grau and Ali Ahmad Jalali, "Forbidden Cross-
Border Vendetta: Spetsnaz Strike into Pakistan during the Soviet-Afghan War,"
Journal of Slavic Military Studies (December 2005): 1.

12.   A. Hakimi, "Nearly 4,000 Families Flee Pakistan for Kunar," *Trend
News*, 4 October 2008, http://en.trend.az/news/politics/1310888.html (accessed
8 February 2011); Sterling L. DeRamus, "Operation Strong Eagle," Unpublished
manuscript, 2010, 1-2; Vowell interview.

13. Vowell interview; Syed Saleem Shahzad, "A Fighter and a Financier," *Asia
Times*, 22 May 2008, http://article.wn.com/view/2008/05/22/A_fighter_and_a_
financier/ (accessed on 29 March 2011); Bill Roggio, "Taliban commander Qari
Zia Rahman Denies Reports of his Death," The Long War Journal, 14 April 2010,
    http://www.longwarjournal.org/archives/2010/04/taliban_commander_qa_1.
php#ixzz1I0pcpFHZ, (accessed on 29 March 2011).

14. DeRamus, "Operation Strong Eagle," 1-2; Vowell interview.

15. Vowell, interview.

16. DeRamus, "Operation Strong Eagle," 2; "Kunar Province Sees Second
Full Day of Successful Combined Operations," Defense Video and Imagery
Distribution System website, 29 June 2010, http://www.dvidshub.net/news/52133/
kunar-province-sees-second-full-day-successful-combined-operations (accessed
on 30 March 2011).

17. Vowell, interview.

18. Vowell, interview.

19. Vowell, interview.

20. DeRamus, "Operation Strong Eagle," 2-3.

21. Captain Steven Weber, First Lieutenant Stephen Tangen, First Lieutenant Douglas Jones and First Lieutenant Spencer Propst, interview by Major Donald Loethan, 4 August 2010; TF No Slack, "Strong Eagle AAR," (unpublished document, 10 July 2010).

22. Weber, Tangen, Jones and Propst interview; Kate Brannen, "M-ATV Production Nears End; Future Contracts Loom," *Defense News*, 22 September 2010, http://www.defensenews.com/story.php?i=4786889&c=AME&s=LAN (accessed on 29 March 2011); PEO Soldier, "CROWS Surge to Afghanistan Along with Troops," *army.mil*, 23 February 2010, http://www.army.mil/-news/2010/02/23/34867-crows-surge-to-afghanistan-along-with-troops/ (accessed on 29 March 2011).

23. "200th Buffalo MRAP Delivered to Military," *UPI.com*, 6 June 2008, http://www.upi.com/Business_News/Security-Industry/2008/06/06/200th_Buffalo_MRAP_delivered_to_military/UPI-15781212771850/ (accessed on 29 March 2011); Weber, Tangen, Jones and Propst interview.

24. Vowell, interview; Weber, Tangen, Jones and Propst interview.

25. "Strong Eagle AAR;"Vowell interview; Weber, Tangen, Jones and Propst interview.

26. Vowell, interview; Weber, Tangen, Jones and Propst interview.

27. Weber, Tangen, Jones and Propst interview.

28. Vowell, interview.

29. Weber, Tangen, Jones and Propst interview.

30. Vowell, interview.

31. Weber, Tangen, Jones and Propst interview.

32. Weber, Tangen, Jones and Propst.

33. Weber, Tangen, Jones and Propst.

34. Weber, Tangen, Jones and Propst.

35. Vowell interview; Weber, Tangen, Jones and Propst interview; DeRamus, "Operation Strong Eagle," 3.

36. Weber, Tangen, Jones and Propst interview.

37. Vowell, interview.

38. DeRamus, "Operation Strong Eagle," 5-6.

39. Weber, Tangen, Jones and Propst interview.

40. Weber, Tangen, Jones and Propst.

41. Weber, Tangen, Jones and Propst.

42. Weber, Tangen, Jones and Propst.

43. Weber, Tangen, Jones and Propst.

44. "STRONG EAGLE AAR."

45. "STRONG EAGLE AAR."

46. Weber, Tangen, Jones and Propst interview; Dianna Khan, "'A Tough Lesson on Who Wants It More,'" *Stars and Stripes*, 21 September 2010, http://www.stripes.com/news/a-tough-lesson-on-who-wants-it-more-1.119045 (accessed on 30 March 2011). LZ Hen was later redesignated OP Thomas after the Soldier killed there.

47. Khan, "A Tough Lesson on Who Wants it More."

48. Weber, Tangen, Jones and Propst.

49. Weber, Tangen, Jones and Propst.

50. Dianna Khan, "'I Don't Know if Discipline is Something You Can Teach,'" *Stars and Stripes*, 21 September 2010, http://www.stripes.com/i-don-t-know-if-discipline-is-something-you-can-teach-1.118919 (accessed on 30 March 2011); Weber, Tangen, Jones and Propst interview.

51. Khan, "'I Don't Know if Discipline is Something You Can Teach;'" Weber, Tangen, Jones and Propst interview.

52. Weber, Tangen, Jones and Propst interview.

53. *JFIRE: MULTI-Service Tactics, Techniques, and Procedures For The Joint Application Of Firepower, FM 3-09.32/ MCRP 3-16.6A/NTTP 3-09.2/ AFTTP(I) 3-2.6*, (Fort Monroe, VA: US Army, 2007), 108, 110.

54. Weber, Tangen, Jones and Propst interview; Khan, "'I Don't Know if Discipline is Something You Can Teach.'"

55. Weber, Tangen, Jones and Propst interview.

56. Vowell, interview; Weber, Tangen, Jones and Propst interview.

57. Weber, Tangen, Jones and Propst interview.

58. Weber, Tangen, Jones and Propst.

59. Vowell, interview; Weber, Tangen, Jones and Propst interview.

60. Weber, Tangen, Jones and Propst interview.

61. Weber, Tangen, Jones and Propst.

62. Khan, "A Tough Lesson on Who Wants it More;" Vowell interview.

63. Albert L. Kelly, "Manning of Outpost Thomas," cjtf-1001.com, 20 July 2010 http://www.cjtf101.com/en/regional-command-east-news-mainmenu-401/3031-manning-of-outpost-thomas-.html (accessed on 13 April 2011); Khan Wali Salarzai, "Residents Support Military Operation in Kunar," *Pajhwok Afghan*

News, 3 July 2010, http://www.pajhwok.com/en/2010/07/03/residents-support-military-operation-kunar (accessed on 30 March 2011); Vowell interview.

64. Khan Wali Salarzai, "Residents Support Military Operation in Kunar," *Pajhwok Afghan News*, 3 July 2010, http://www.pajhwok.com/en/2010/07/03/residents-support-military-operation-kunar (accessed on 30 March 2011); Vowell interview. "Gates Awards Bastogne Valor," *Fort Campbell Courier*, 9 December 2010, http://www.fortcampbellcourier.com/news/article_98b9208c-03da-11e0-9dec-001cc4c03286.html (accessed on 4 April 2011); "'No Slack' Soldiers Fight Relentlessly, Repel Insurgent Attacks," *Bagram Media Center*, 7 January 2011, http://www.clarksvilleonline.com/2011/01/07/no-slack-Soldiers-fight-relentlessly-repel-insurgent-attacks/ (accessed 4 April 2011). The awardees were: Silver Star, CPT Steven Weber, 1LT Steven Tangen, SFC John Fleming, SSG Brent Schneider, SSG Daniel Hayes, CPL Joshua Busch, PFC Richard Bennett; Bronze Star, 1LT Douglas Jones, SFC John T. Howerton , CPL Joshua M. Frappier.

65. Vowell, interview.

66. Vowell, interview; "Task Force No Slack Continues Clearing the Ghakhi Valley," *Clarksville Online*, 1 August 2010, http://www.clarksvilleonline.com/2010/08/01/task-force-no-slack-continues-clearing-the-ghaki-valley/#more-42212 (accessed on 30 March 2011).

67. Albert L. Kelley, "Marawara District Shura After Strong Eagle II,"cjtf82.com, 21 July 2010, http://www.cjtf82.com/en/regional-command-east-news-mainmenu-401/3034-marawara-district-shura-after-operation-strong-eagle-ii.html (accessed on 30 March 2011); Khan, 'A Tough Lesson on Who Wants it More;'" Bill Roggio, "Afghan, Us Forces Hunt Al Qaeda, Taliban in Northeast," Long War Journal, 2 August 2010, http://www.longwarjournal.org/archives/2010/08/afghan_us_forces_hun.php#ixzz1IYYFTbiv (accessed 4 April 2011).

68. Praveen Swami "Linda Norgrove: US Forces Hunting Down Kidnap Group," *The Telegram* [London], 13 October 2011, http://www.telegraph.co.uk/news/worldnews/asia/afghanistan/8061282/Linda-Norgrove-US-forces-hunting-down-kidnap-group.html (accessed 4 April 2011).

69. Vowell, interview.

70. "STRONG EAGLE AAR."

71. Weber, Tangen, Jones and Propst interview.

72. Vowell; Weber, Tangen, Jones and Propst.

73. Vowell, interview.

74. Vowell, interview.

75. Combined Joint Task Force 101, "Joint Forces Continue Marawara Operations," Defense Video and Imagery Distribution System website, 3 April 2011, http://www.dvidshub.net/news/68199/joint-forces-continue-marawara-

operations (accessed on 4 April 2011*); Scott Rottinghaus, "Wednesday Warfighter: Progress and Pitfalls in Marawara," *DOD Live*, 30 March 2011, http://www. dodlive.mil/index.php/2011/03/wednesday-warfighter-progress-and-pitfalls-in-marawara/# (accessed 4 April 2011).

Disrupt and Destroy
Platoon Patrol in Zhari District, September 2010
by
Matt M. Matthews

On the morning of 5 September 2010, 1st Platoon, Headquarters and Headquarters Troop, 1st Squadron, 75th Cavalry Regiment (1/HHT/1-75 CAV) of the US Army's 2d Brigade Combat Team (2BCT), 101st Airborne Division (Air Assault) embarked on its first combat patrol south of Highway 1 in the Zhari District, Kandahar Province. The mission would take 1st Platoon into the heart of Taliban territory as part of the initial 1-75 CAV effort to dismantle the enemy infrastructure and force its withdrawal from Zhari. As the HHT commander pointed out afterwards, "The fight in Tiranan set the tone for the rest of the year and let the Taliban know that the unit they were fighting would not withdraw under sporadic or even heavy fire." To be sure, the enemy was taken aback by the aggressive character of the operation and immediately realized that its Zhari stronghold was not impenetrable. The mission proved a learning experience for the platoon leader and his young Soldiers and demonstrated what a well led, highly motivated, extremely competent platoon can accomplish against an entrenched and determined foe.[1]

Photo courtesy of Captain Riley E. McEvoy

Figure 1. First Platoon embarking on its first combat patrol south of Highway 1 in the Zhari District, Kandahar Province.

Background

By the early summer of 2010, a resurgent Taliban had enveloped Kandahar Province. Across large swaths of the province, explosions and billowing plumes of smoke revealed the presence of a determined and resilient enemy. Indeed, by the beginning of the spring fighting season of 2010, the Taliban had assembled a formidable force around Kandahar City. In a series of offensive actions dating back to 2003, they had captured key terrain around the city which allowed them to intimidate the population and launch attacks against the International Security and Assistance Force (ISAF) and Afghan security forces. A relative paucity of Soldiers in Kandahar Province during the period 2003-2010, significantly restricted the efforts of ISAF and its Afghan partners who had endeavored for years to destroy the Taliban in the Province.[2]

As the Taliban increased its attacks in the province, ISAF in early 2010, prepared to strike back with a hard hitting and wide ranging counteroffensive to drive the Taliban away from the city. This action, made possible by the recent troop surge, aimed to strike directly into several crucial districts in Kandahar Province. One of the targeted districts, Zhari, dubbed the "heart of darkness" by apprehensive Soviet troops sent there in the 1980s, was also the birthplace of the Taliban. One specialist in the field pointed out the importance of the operation when she stated, "To strike at the heart of the insurgency, strike at the historical and spiritual home of the Taliban sends a very clear message – with the resources we have, we are on the offensive."[3]

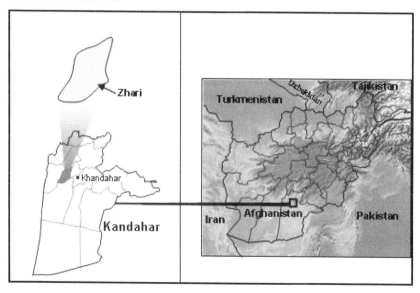

Figure 2. Kandahar Province, Zhari District.

By 1 September 2010, two infantry battalions and one cavalry squadron of 2BCT, 101st Airborne Division (Air Assault) had occupied positions in Zhari along Highway 1. The 1-75 CAV occupied 2BCT's center sector moving into Forward Operating Base (FOB) Wilson in early September. Under the command of Lieutenant Colonel Thomas McFadyen, the unit was responsible for a sizable portion of Zhari. As a Reconnaissance, Surveillance and Target Acquisition (RSTA) unit, 1-75 CAV was not as large as the infantry battalions in 2BCT and its Modified Table of Organization and Equipment (MTOE) differed greatly. Alpha and Bravo Troops of 1-75 CAV were, for the most part, made up of cavalry scouts while Charlie Company (also known as "Chaos" Company), was an infantry element.

Prior to deploying to Afghanistan, Chaos Company's 1st Platoon trained with HHT. The Platoon Leader, First Lieutenant Riley E. McEvoy, wrote that, "During JRTC [Joint Readiness Training Center] they made the decision that HHT needed a maneuver element." Halfway through their training at JRTC, McEvoy's platoon was assigned to HHT but at the end of the training program they were sent back to their own company. Within weeks of the Squadron's arrival in Afghanistan however, 1st Platoon was reassigned to HHT and given the unit designation 1st Platoon, Headquarters and Headquarters Troop, 1st Squadron, 75th Cavalry (1/HHT/1-75 CAV).[4]

There were two important reasons for attaching McEvoy's Platoon to HHT. The first object was the need for four maneuver companies as partners with the Afghan National Army (ANA) Kandak (Afghan Battalion) which was attached to the 1-75 CAV. There were four ANA companies and without the attachment of 1st Platoon to HHT there would be only three maneuver elements to partner with the Kandak. First Platoon was also assigned to HHT because the Squadron Commander wanted a maneuver force which could work with his Intelligence Officer (S2). McEvoy recalled that McFadyen sought "to have a maneuver force partially controlled by the Squadron S2 and himself to action high-value intelligence targets and time-sensitive targets." The S2, Captain Matthew A. Crawford, described 1st Platoon's mission as "unique." They were used "to specifically disrupt, target HVIs (high-value individuals), and destroy known insurgent infrastructure. Their mission statement almost always had the words 'disrupt' and 'destroy' in them." McEvoy used few words to explain why his platoon was chosen, simply stating, "We were the best."[5]

McEvoy's Platoon contained some of the most experienced Non-Commissioned Officers (NCOs) and enlisted men in the US Army. His Platoon Sergeant, Sergeant First Class Derek Leach, was a solid combat veteran and a superb leader, so much so that McEvoy pointed out their relationship "was perfect." The Platoon Leader wrote that Leach, "brought a fresh attitude and perspective to the platoon. He was…gung-ho and ready to train and fight." The Squad Leaders, Staff Sergeant Timothy McKinnis and Sergeant Victor Faggiano, were also first rate combat veterans. McEvoy stated that McKinnis was "just a wealth of experience, [an] amazing NCO, and it was a great mix between Sergeant Faggiano …a young fire breather, ready to go at all times and then Sergeant McKinnis… [an] old crusty NCO that has so much perspective about everyone in the squad, and just so much experience and intelligence to share, and just a great, calm combat leader." McKinnis' 2d Squad also contained two excellent Team Leaders, Sergeant Jesse Hattesohl and Sergeant Zachary Fraker. In Faggiano's 1st Squad, Team Leaders Specialist Joseph Lee and Specialist Cody Chandler imparted a wealth of knowledge, skill, and leadership. Chandler was also one of the most popular Soldiers in the platoon. McEvoy's weapons squad was led by Staff Sergeant Robert Singley who had just returned from drill sergeant duty. Like the others, he was a solid combat veteran and leader.[6]

The Mission

By 1 September 2010, Crawford had produced intelligence packages on numerous enemy compounds, fighting positions, meeting locations, and bed-down locations. He had also accumulated a massive amount of intelligence material on Taliban Improvised Explosive Device (IED) caches and IED production facilities. All of the sites had been identified either by historical reporting or other intelligence, surveillance, and reconnaissance (ISR) means. With the addition of full-motion video shot by Unmanned Aerial Vehicles (UAVs), it did not take Crawford long to indentify certain enemy patterns. "So we had a…patterned enemy in Zhari…he was so comfortable with the terrain and the fact that they owned it that they just…became almost lazy in their movement patterns," the S2 remarked.[7]

The Squadron's initial objective was to force the enemy away from Highway 1 and Route SUMMIT in order to set conditions for an impending 2BCT operation dubbed Operation DRAGON STRIKE. "We wanted to shape the AO, so once DRAGON STRIKE started we could go in and not have to slug it out," Crawford recounted. The entire operation was intelligence-driven, targeting Taliban compounds, cache sites, and bed-down locations near Highway 1 and Route SUMMIT. "We wanted to open up those two major routes in our AO…and we were running patrols 750 meters in, 1,000 meters in, making contact, and then trying to kill the enemy with artillery and UAV strikes," Crawford recalled.[8]

Early in the first week of September, McEvoy received orders to prepare his platoon for a mission south of Highway 1. The operation would take place on the morning of 5 September. Although they had performed combat patrols north of the highway, 1st Platoon had yet to conduct operations to the south. It would be an extremely dangerous mission. The Taliban owned the ground, and the terrain which was a jumble of dense grape fields, *wadis*, tree lines, and open grassland and this only served to increase the strength of the enemy. As Crawford pointed out, "you couldn't get 300 meters south of that highway without getting attacked." Crawford wanted McEvoy's platoon to check a possible IED cache site west of Route SUMMIT. The S2 was also convinced that the enemy was using at least one building along Route ST. JOHN'S, a dirt road running north of Tiranan, as a Command and Control (C2) point.[9]

As soon as they received the warning order, McEvoy and his NCOs began their pre-combat checks (PCCs) and pre-combat inspections (PCIs). They received reliable intelligence on possible routes and the locations of recent IED attacks. "We actually got really deep into finding intelligence about this area," Hattesohl recalled. First Platoon would be supported for this mission by UAVs, 155mm howitzers, and 120mm mortars from FOB Wilson and Close Combat Attack (CCA) aircraft. With all necessary preparations in place, McEvoy's platoon and their ANA counterparts were fully prepared by the night of 4 September.[10]

The Patrol

At approximately 0700 on 5 September, 1st Platoon and the ANA platoon started moving south from FOB Wilson. The patrol should have moved out earlier but the ANA contingent arrived late. Hattesohl, the 2d Squad A Team Leader, led the dismounted platoon through the Afghan National Police (ANP) gate. Following behind Hattesohl were his two team members, Private First Class Steven Craythorn and Private First Class Joseph Arvizu. Hattesohl's team advanced south toward Highway 1 in a staggered road-march formation. They were followed by about half of the ANA platoon along with 2d Squad Leader McKinnis and Fraker, the B Team Leader. Fraker was the lone man in B Team, as two of his Soldiers, Specialist Ryan Johnson and Private First Class Yeng Her, were on leave.[11]

Toward the center of the column, McEvoy maintained a watchful eye on the balance of the ANA intermingled with the rest of the patrol. "It was one of our first patrols," McEvoy recalled, "so controlling the ANA was a bit difficult." With the platoon leader was Captain Michael Krayer, the HHT Commander, who had come along as an observer and Private First

Class Mark Drake, the forward observer (FO). Private First Class Anthony Surrett also traveled with McEvoy serving as his radio telephone operator (RTO).

Behind this element, the 1st Squad A Team Leader, Chandler, advanced with Private Michael Iacoviello and Private First Class Kyle Hall. In the center, 1st Squad Leader Faggiano and attached HHT Medic Sergeant Timothy "Doc" Peterson moved forward. Behind Faggiano, 1st Squad B Team Leader Lee and his two Soldiers, Private First Class Anthony Thompson and Private First Class Hector Bonilla, followed their squad leader through the front gate. Also moving with Faggiano was Weapons Squad Leader Singley and his M240G gun team members, Private First Class Spencer Harris, Private First Class Rigoberto Soto and Private First Class Corey Doty.[12]

Bringing up the rear of the patrol was Platoon Sergeant Leach, Medic Private First Class Robert Kruse and the designated marksman and ammo bearer (AB) from Weapons Squad, Specialist Andrew Ingram. Ingram carried an "assault pack" of 500 rounds of 7.62 ammunition for the gun team and he also carried an M14 rifle. As the platoon began to stretch out between the front gate and Highway 1, Leach and Kruse, along with Ingram, began moving toward the front of the long column. "As usual," Leach stated, "we would patrol up and down between everyone ensuring everyone was good and also for my situational awareness and coordination with the platoon leader and squad leaders." As his platoon marched through the ANP gate, McEvoy called the command post (CP) to announce their departure. He remembered the day as "hot, over 100 [degrees], sunny [and] cloudless."[13]

At Highway 1, 1st Platoon set up left and right security, stopping all traffic and quickly moving across to the south side of the highway. Once there, the platoon changed from a staggered road-march formation into a squad wedge formation and headed west through an open field toward Route SUMMIT. Hattesohl remained on point during this phase of the movement. Crossing Route SUMMIT McEvoy's Soldiers cautiously approached the rural community of Pasab. "We continued with this movement until we made it closer to the village of Pasab where we set in security watching every avenue of approach and especially Highway 1," Leach recalled. Ever mindful of the threat from IEDs, Hattesohl continued to select the "hardest and safest route possible" for the platoon. Moving north of Pasab, Leach remembered that, "Once we started moving between buildings we switched to a bounding and traveling overwatch [formation]

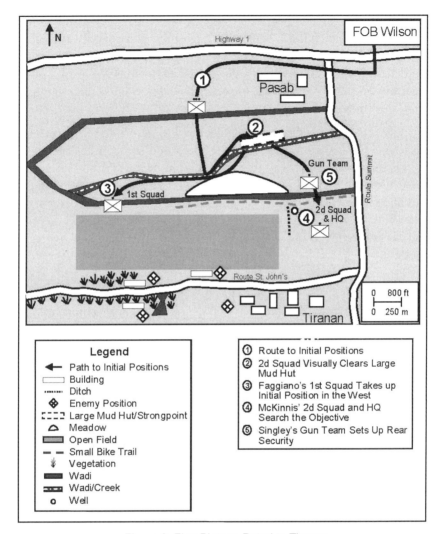

Figure 3. First Platoon Patrol to Tiranan.

to ensure security was set with one element while still being able to push on with the other element." McEvoy stated that he gave few instructions up to this point. "The squad leaders knew their route and what they were supposed to do."[14]

The patrol moved west past Pasab about 600 meters and then turned south toward the first *wadi* line. When Hattesohl approached the *wadi*, he called a halt to check for "danger zones." He recalled that there was little water in the *wadi* but there were "a lot of trees a lot of bushes" and "You couldn't see more than maybe 15 [or] 10 feet in front of you." After the brief pause, the platoon began crossing the riverbed. "For this mission we came up to the *wadi* [and] walked through the water roughly 100 meters or so," Leach stated. "For this movement we were in single file, but spaced well enough apart and mutually supporting each other. We would push a team up, find the crossing, set in security while bringing up a gun team, [and] then move everyone else up."[15]

While the rest of the platoon made its way across the first *wadi* line, Hattesohl and his team moved farther south into an open field about 100 meters long. About halfway through the field, Hattesohl noticed a small piece of plastic. Believing it could be an IED the team leader called another halt and moved to clear it. "Everything's a hazard for us," he later stated. After checking the piece of plastic Hattesohl declared, "It was nothing," and his team moved out once more in the direction of a tree line to the south.[16]

Making their way past the tree line, Hattesohl's team proceeded about 200 meters to the next *wadi* line. The team leader described this obstacle as a "creek *wadi*." Entering the creek, McKinnis' 2d Squad, with Hattesohl's team on point, moved northeast toward a large mud hut with an oval roof. While 2d Squad moved on the structure, Faggiano's 1st Squad set the gun teams to provide security for the other squad. "We had learned from [past experience] the terrain is so dense that a machine gun [placed] in [a] grape orchard can't really cover anybody," Faggiano noted. The gun teams and Faggiano's 1st Squad provided overwatch for McKinnis' 2d Squad as they approached a large mud hut with a small open courtyard surrounded by damaged walls. As 1st Squad moved up to the structure, Faggiano's Soldiers and the gun crews continued to provide overwatch. McKinnis' squad found nothing at the house. McEvoy recalled "visually" cleaning the house. "Nothing was really there," Hattesohl reported. "Our leadership talked with the locals that were living in that village…we asked them questions like where have the Taliban been…has there been any activity in the area?" It was all for naught. The platoon received no actionable intelligence from the local population. Finding neither IED caches nor IED production facilities in the vicinity, McEvoy made the decision to move south to the objective.[17]

Once again, Hattesohl led the way for McKinnis' 2d Squad, south toward the third *wadi* line and the objective. At the same time, Faggiano's 1st Squad headed in a southwesterly direction toward the third *wadi* line. Making their way through the dense underbrush, Faggiano's Soldiers emerged on the south side of the *wadi*. "Just south of me, it was wide open, there was tall grass, and then a village just off to the south," the 1st Squad

Leader recalled. Faggiano's squad quickly set up an overwatch position to cover 2d Squad as they worked their way toward the third *wadi* line. Lee and his team faced north to provide security for 2d Squad. McKinnis stated that 1st Squad's mission was "to overwatch movement and deny the enemy use of the wadi line to the west while my squad went to the objective to confirm or deny it as a cache site."[18]

After working their way through the *wadi*, McKinnis' men emerged onto an open field near the objective, which McEvoy remarked, "turned out to be a well." With 2d Squad were McEvoy, the FO Drake, Leach, Krayer, his security element, as well as 10 ANA Soldiers and the gun teams. A quick search of the well turned up nothing. "[The] primary objective was the well, which was believed to have been a cache site," McKinnis recalled. "After discovering that the objective was not a cache site, we moved about 50-75 meters further southwest to the northern corner of the grape field in order to set a small perimeter." Hattesohl and his A Team, along with Harris and his M240G from the gun team, took up an overwatch position in the southwest corner behind a wall near the grape field. Fraker also took up a position behind the wall. Other Soldiers in the element pulled security to the northwest and east. Hattesohl was soon joined by McKinnis who crouched behind the wall. While this was taking place, the other gun team which included Singley, Doty and Soto, set up rear security on the north side of the *wadi*.[19]

The Fight

As McKinnis' men took up their positions, an ANA Soldier fired a warning shot at an Afghan "farmer" in the grape field. The farmer turned and fled whereupon several ANA Soldiers bolted into the grape field after him. At this point, McEvoy decided to wait and to keep pulling security until the ANA returned from pursuing the man. The ANA soon led the farmer out of the grape field. Hattesohl remembered talking to the man, tactfully questioning him about the Taliban presence in the area. Seated nearby on a haystack, McEvoy recalled McKinnis and the ANA chatting with the farmer. To the west, Faggiano stated that his squad was "kind of obsessing over what was going on" in the 2d Squad area.[20]

About five to eight minutes into the questioning of the farmer, Hattesohl heard another gunshot. "I heard a snap. I thought it was one of the ANAs shooting again but it sounded a little bit different. Then it was on. It was game on from there. I called the direction and distance. I saw that there were two lone buildings just to my southwest, about…300, 350 meters and that's where we were taking the fire from." Fraker also heard the gunfire from behind the wall north of the grape field. "We heard a couple pop shots and then everything broke loose," he stated. From their rear security position north of the third *wadi*, Singley, Soto, and Doty could hear the rounds going over their heads. Kruse recalled that the fire was so intense that "we were pinned down." He remembered being exposed to the initial enemy fire and sprinted to the wall north of the grape field under a hail of bullets. "There [were] rounds hitting the grape field in front of us [and] behind us," he later stated. Although under a storm of small arms fire, 1st Squad, the gun team, and the headquarters element struck back swiftly against the enemy. Hattesohl and Harris with his M240G were the first to return fire from north of the grape field.[21]

Sergeant Fraker could see about 20 enemy fighters in and around two buildings to his southwest. He recognized instantly that the squad's weapons could not fire effectively on the Taliban locations. Under a withering barrage of bullets from enemy small arms and machine guns, Fraker and Hattesohl moved from their defensive position behind the wall about 25 meters into an open field. As they moved forward, other members of the squad laid down a suppressive volley with their M4s and M249Gs. Once in the open field Hattesohl put down suppressive fire while Fraker fired his M320 at the enemy to his front. As Fraker's 40mm high explosive grenades burst around the Taliban defensive positions, Hattesohl continued to direct his team and coordinate its fires.[22]

To the west, Faggiano's 1st Squad also came under a torrent of gun fire. He immediately called up a contact report to McEvoy and began to study the situation. "There was really nowhere to maneuver," he later noted. "These guys were the masters of their terrain. They knew we weren't going to come at them. They knew we were going to have to slug it out with them, from wadi line to wadi line." Faggiano quickly discerned that he needed to gain fire superiority over the enemy and began coordinating with his team leaders Chandler and Lee.[23]

Most of the fire was coming from the *wadi* line directly to the south of 1st Squad and from two buildings on the north side of Route ST. JOHNS. Faggiano recalled listening to McEvoy on the radio. The platoon leader had already called in a contact report to the troop command post (CP). He thereupon assisted the FO Drake in coordinating his radio calls to the squadron for Close Combat Attack (CCA) aircraft (OH-58 Kiowas), along with indirect fire support. For now, McEvoy was determined to let his squad leaders "fight for a while" until he could acquire more air and artillery. Faggiano remembered McEvoy saying, "All right, we're going to get CCA. Let's hold up right here." Bolstered with the knowledge of the enemy's locations, McEvoy determined to bring air and artillery down on them while his squads attempted to gain fire superiority.[24]

As Faggiano's 1st Squad fired back, they immediately drew added attention from the enemy to their south. "They obviously were the ones that the Taliban saw first so they took most of the rounds," Hattesohl noted after the fight. "Then when 1st squad opened up, that's when the enemy shifted fire towards them." Indeed, Faggiano's squad was engulfed in a hail of small arms fire which whizzed over their heads, ripped into the dirt and sliced through the vegetation along the *wadi* line. The enemy fire was so intense that Faggiano was not certain he could maintain fire superiority. The squad leader now called on Lee, who with Thompson and Bonilla, had been covering the rear on the north side of the *wadi*. "I wanted to bring more [M]320's and team leaders…I wanted to bring more of their gunners over, so Lee jumped over and we left two guys facing north," Faggiano stated.[25]

Although the enemy was outside the range of the M320, Lee recalled firing HE rounds as a scare tactic. On the east side of Faggiano's line, Chandler, Iacoviello, and Hall laid down a steady stream of fire with their M203s, M320s, and the SAW. Chandler could see large groups of the enemy running in and out of a building directly south of his position. As the enemy ran out of the structure, they would try to establish firing positions and then duck back into the house. As the firefight grew in intensity, Lee began rotating his team from rear security to the firing line in order to conserve ammunition. To Lee's great consternation, the enemy continued their withering assault on 1st Squad. "They weren't trying to bound back and leave. They were actually sticking there and fighting with us. They just had all the buildings that they could hide in where we had a berm with trees and basically a couple walls, nothing major to really stop anything," Lee remembered.[26]

In the east, 2d Squad continued to fire back at the buildings to the southwest. From his position in the open field, Fraker noticed about 10 enemy fighters attempting to move east along Route ST. JOHNS. Realizing the Taliban was attempting to flank 2d Squad's position, Fraker instantly reoriented a portion of the squad's fire on the enemy's flanking party. The intense gunfire quickly halted the flanking movement. While this was taking place, Hattesohl continued to move back and forth between the open field and his team's defensive position. Hattesohl alternated between these two positions in order to engage the enemy with the squad's highest-caliber weapons. He also continued to provide his team with tactical instructions and direct their fires, as the tempo of the fight increased.[27]

Remarkably, a few of the ANA Soldiers at McEvoy's location were wandering around approximately 30 meters in front of 2d Squad's fighting positions. With bullets flying everywhere, the platoon leader promptly ordered an ANA officer to take all of his men back to a small bike trail about 50 meters north of 2d Squad's position. By this time, Drake had already fired several indirect fire missions on enemy positions to the south. As he called in these missions, CCA arrived overhead. While Drake coordinated with the OH-58 Kiowas, McEvoy informed McKinnis that he was concerned about the proximity of the artillery fire and that he planned to pull the platoon back to the bike trail. McEvoy recalled that this was their first real combat patrol and he was greatly concerned about having enough stand-off room. He remembered that during later deployments they called artillery in with less stand-off range, "but this time...we backed off a full 100 meters, which is way more than we should."[28]

As the platoon leader continued to coordinate the planned movement, Faggiano's squad faced new tribulations. The struggle continued to rage on the 1st Squad firing line. Through the heat and uninterrupted enemy fire, Faggiano's squad fought back with skill and determination. The squad leader worked his way up and down the line, repeatedly exposing him to enemy gunfire as he directed his squad's defenses. As 1st Squad fired back at the enemy to their south, Faggiano noticed a slackening in the fire from Lee's team. Bonilla's SAW had malfunctioned and he was unable to correct it. Faggiano rushed to Bonilla's position and attempted to fix it. After feverishly attempting to correct the problem with the SAW, the squad leader discovered that the firing pin was broken. Scrambling back down the *wadi* line, Faggiano learned that Thompson's M203 had also malfunctioned, the result of a broken plunger. Although confronted by this adversity, 1st Squad remained calm and continued to fight back. "Two weapons systems down and they were completely inexperienced at the time, completely inexperienced in the sense of a firefight...they did amazing...everyone kept their cool," Faggiano recalled.[29]

As 1st Squad took stock of its situation, their adversaries maintained a deadly barrage of machine-gun and small-arms fire. In spite of everything, 1st Squad kept up a blistering fire on the enemy. On the left, Chandler, Iacoviello, and Hall fought back with a vengeance while on Faggiano's right, Lee, Thompson, and Bonilla put up the same determined resistance. Looking down the line and shouting out fire commands, Faggiano saw an enemy bullet bounce off the ground and fly right between Thompson's arms as he fired his M-4. The squad leader watched in amazement as Thompson momentarily slid back behind cover and then quickly popped back up to engage the enemy. "They did a great job," Faggiano stated afterward.[30]

Both Faggiano and Chandler could still see a large group of fighters running in and out of the building southwest of their position. Neither Soldier knew it at the time but Captain Crawford, the Squadron S2, had already confirmed that the structure was in fact, a hardened Taliban fighting position and a command and control point. Crawford had amassed data on the building dating back to 2007. Now, with one Predator UAV and one Reaper UAV flying overhead, along with other fixed-wing aviation assets, the TOC could clearly see numerous 5-10 man enemy elements moving forward from south of Tiranan and Route ST. JOHNS. The fighters were clearly reinforcements.[31]

The building was certainly a lucrative target and Chandler yelled to Faggiano, "Hey, see where they're shooting from?" The squad leader saw the fire coming from the building and shouted over the deafening gunfire, "Go and shoot the SMAW-D [Shoulder Launched Multipurpose Assault Weapon-Disposable]. Wreck that building!" Chandler immediately clutched the SMAW-D, brought the weapon to his shoulder, aimed and fired. As the round careened toward the building, Chandler thought the back blast from the weapon would clearly expose his position. Unfortunately the round missed the target. Turning to grab his M203, he continued to fire HE rounds toward the enemy. As Faggiano started to move down the line in order to help Lee spread his team out along the brim of the *wadi*, the squad leader turned to tell Chandler where he was going just as the team leader closed the breech on his M203. At that moment, a bullet slammed into Chandler's left arm as another struck his lower abdomen. "It sounded like an explosion," he recalled. The blow knocked the young team leader down into the *wadi*. "I just remember trying to climb back up to the berm to get my weapon and to get back to where I was at but my legs just gave out on me and I collapsed into the *wadi*."[32]

In an instant, "Doc" Peterson, the headquarters medic, ran down into the *wadi* to help Chandler. Lee quickly followed. Faggiano, the 1st squad leader, was next on the scene. Doc told the squad leader that one bullet had gone through Chandler's arm and the other through his lower abdomen. As Peterson applied pressure on the wounds, he told Faggiano that Chandler was "urgent surgical" and gave the squad leader a hasty 9-Line MEDEVAC request. Wasting no time, the squad leader called the report up to Leach. "Chan's shot. Urgent surgical," he told the Platoon Sergeant.[33]

McEvoy heard the MEDEVAC call and directed Leach to handle it. McEvoy recalled later that the order was unnecessary because Leach and the platoon medic, Kruse, were already in motion. Running through a gauntlet of small arms fire, Leach and Kruse hurriedly made their way to Chandler. "We had to run through the field to the tree line," Kruse noted. When they reached Chandler, Kruse remembered a torrent of enemy gunfire going over their heads. He could also hear the bullets hitting the trees around them.[34]

With Leach, Peterson, and Kruse treating Chandler, Faggiano and Lee returned to the fight. With two weapons systems down and one team leader out of action, the squad leader knew he needed every weapon on the firing line. Climbing back up to the berm, Faggiano saw Iacoviello firing away at the enemy. Iacoviello asked Faggiano about Chandler and then "turned and started shooting again," the squad leader stated. Enemy bullets continued to hit the trees and dirt around them but Faggiano's squad fought back with a renewed vigor.[35]

With the MEDEVAC helicopter on its way, McEvoy grew concerned about the amount of enemy fire on 1st Squad's position. The MEDEVAC would be forced to land in a large meadow just behind their location. As the helicopter approached, the platoon leader told Leach to mark the landing zone (LZ) with smoke and directed his Soldiers to lay down a sheet of fire as soon as the Black Hawk came into view. "I just need everyone to lay down cycling, just pretty much hold the trigger down toward the enemy so they keep their heads down while we evac[uated] Specialist Chandler," McEvoy reported. With the MEDEVAC on its way, Drake had the artillery stop firing. "I had the guns go cold as I was using the birds," he later pointed out.[36]

As the MEDEVAC approached the LZ, 1st and 2d Squads, along with the weapons teams, laid down a sheet of fire on the enemy positions to the south. However, as the Black Hawk set down in the meadow, the enemy fire increased. Lee noted that, "They picked up the rate of fire and so did we." When the Black Hawk landed, enemy rounds smacked into the trees and several bullets kicked dirt into Thompson's face. Lee recalled that 1st Squad "was pretty much pinned down." Chandler stated afterward that, "once the bird started coming in, all hell just broke loose. The Taliban just started shooting at the bird." As soon as the MEDEVAC landed, the crew chief jumped off and helped "Doc" Peterson, Leach, Kruse, Hall, and Lee carry Chandler to the waiting aircraft. While placing him in the Black Hawk, enemy rounds cut through the fuselage. "They loaded me into the bird and as we were taking off, I just started hearing bullets hit the bird and at that point, I just thought, I'm going down. We are going down and I was actually very surprised that they even landed that bird where they did." Chandler recalled.[37]

With the MEDEVAC airborne and heading north, McEvoy could now pull 2d Squad back to the bike trail and call for more artillery fire. Drake and McEvoy called in 155mm howitzers and 120mm mortars on enemy locations north of Route ST. JOHNS as McKinnis' Squad and part of the gun team hurried north to the bike trail. As the rounds exploded around the Taliban positions, 2d Squad took up positions along the bike trail with their ANA counterparts. "We pulled back [and] called for rounds," Hattesohl stated. He remembered the bike trail as "a very small ditch," that "provided…little cover but the concealment was great, there was lots of vegetation, there was no way they could have seen us." While McKinnis' Soldiers took up their new positions along the trail, Singley's other gun team continued to provide rear security on the north side of the *wadi*. Enemy fire slackened as 2d Squad secured its position along the bike trail. "The rounds weren't coming toward us anymore, they were all going towards 1st squad," Hattesohl pointed out.[38]

Around the enemy-occupied buildings, howitzer and mortar rounds landed with thunderous flashes. At the same time, Kiowas returned to the area searching for targets. Drake called a halt to the indirect fire and the platoon waited for the helicopters to strike. Armed with rockets and .50-caliber machine guns, the Kiowas were prepared to attack. However, they were not sure of the Taliban's precise locations and needed someone to mark the targets. Hattesohl recalled, "Kiowas having a really tough time finding the enemy position." Although 2d Squad was able to see the enemy locations from the bike trail, their M320s and M203s were about 150 meters out of range. Hattesohl turned to Fraker and told him that they would have to go and mark the target. Rushing to McKinnis' position, Hattesohl asked his squad leader for permission to mark the target with smoke with the M203 grenade launcher and said he would be taking Fraker with him. The request was quickly granted. As Hattesohl and Fraker ran south toward their former position, McKinnis ordered the rest of his squad to lay down cover fire.[39]

Running about 250 meters into an open field in front of the bike trail, Hattesohl and Fraker found themselves in a freshly tilled field of black dirt. "I laid down, Sergeant Fraker shot the smoke, and I was just shooting explosive grenades," recalled Hattesohl. The two team leaders were quickly "pinpointed" by the enemy and bullets immediately flew around their heads and into the dirt. After firing two or three smoke grenades with the M203, Hattesohl and Fraker fell back to the bike trail. Unfortunately, the smoke quickly dissipated. Fraker later pointed out that M203 "smoke doesn't stay up long." As the Kiowas came back around they told the Soldiers on the ground "we don't see it yet, mark it again." Once again, the two Soldiers ran back into the open field under enemy fire. Fraker fired again and this time the helicopters identified the smoke and began engaging the targets. Fraker recalled the Kiowa's repeatedly "laying waste to that area."[40]

As the Kiowa's continued their gun runs, McEvoy and Drake ran from the safety of the bike trail into the open field to place a VS-17 marker panel to help the pilots identify 1st Platoon's location. While they rushed forward under a barrage of enemy bullets, 2d Squad laid down a blistering cover fire. Placing the marker panel at about the spot from which Hattesohl and Fraker had fired their smoke grenades, McEvoy and Drake hurriedly moved back toward the trail as enemy bullets zipped past their heads. "I remember him [McEvoy] calling ... 'All right; were getting ready to come in,'" McKinnis stated. With 2d Squad laying down heavy cover fire McKinnis recollected, "He didn't want to get shot." In a display of conspicuous bravery and leadership, McKinnis also ran out into the open field under Taliban fire to lay down another marker panel. With the smoke and marker panels in place, the Kiowas blasted the enemy positions. "It was really a release of pent up tension when they started hitting them... there were some cheers going on," McKinnis remarked.[41]

At this point in the fight, McEvoy decided to pull back farther north to a better defensive position. The entire platoon would re-cross the *wadi* and take up positions in a nearby meadow that, surrounded by a wall, appeared to be a rock solid location. McKinnis remembered the meadow had "a three or four foot wall around it where we could make a pretty good [defensive] position while they dropped the rounds." Second Squad and the gun teams pulled back first. They were followed by Faggiano's 1st Squad that sprinted in from the west under the cover of Kiowas. For the first time in hours, the entire platoon was consolidated at the same location in a 360 degree defensive position.[42]

Although the new position was first rate, the platoon was still under fire. Hattesohl said the enemy rounds coming closer and the noise getting louder. He also noticed that the Taliban's fire was becoming much more accurate. Soon, enemy grenades started landing close to the platoon's position. Indeed, the enemy's familiarity with the terrain had allowed a few fighters to press in closer to 1st Platoon's new location.[43]

Drake soon realized that he could not properly call for or adjust fires from the new location and requested permission to move forward to the *wadi*. On receiving the go ahead from McEvoy, Drake, along with Faggiano, Hall, and Iacoviello, scrambled back to the *wadi* line. Although enemy fighters were clearly close by, Faggiano recalled Hall jumping right into the *wadi* line, "ready to fight those guys wherever they were and they were just right on the other side." As they moved forward, McKinnis lobbed several grenades over to the southern side of the *wadi*. A few seconds later, Faggiano heard an enemy fighter close by. "I think he had crept up and was just throwing pot shots, trying to follow us down the *wadi* line," he later stated. The squad leader quickly tossed a grenade at the man and heard nothing more from him.[44]

In the thick underbrush along the *wadi*, Drake could not find a suitable position to determine targets and told Faggiano he couldn't see from that location. "This isn't a good spot," he said. With Drake unable to see and with more enemy fighters pressing in on their flanks, Faggiano decided to fall back to the meadow. As they moved back north, McEvoy decided to move the entire platoon out of the meadow. He would move further north to the little mud hut the platoon had searched that morning and turn it into a strongpoint. The new position would also provide the FO with a better view of the enemy positions to the south.[45]

"We all peeled back to the house," Faggiano said. While his squad set up security on the western side of the building, McKinnis' 2d Squad took up positions to the east and south. "It was a mud hut and it was actually pretty big," Hattesohl remembered. He said a family had been living in the hut with their dog. Wasting little time, McEvoy placed Ingram with his M-14 and Doty, Harris, and Soto with their M240Gs on the roof along with Drake. Faggiano remained close by to provide command and control. Using his 10-power scope on his M-14, Ingram could clearly see Taliban fighters firing from numerous locations about 700-800 meters to his south. He also saw enemy reinforcements rushing forward south of Route ST. JOHNS. He quickly relayed the information on the enemy targets to McEvoy, Drake, and the gun teams. While Doty and Harris fired their M240Gs at enemy targets, Drake began calling in Kiowa and 155mm howitzer missions.[46]

Much of the enemy activity seemed centered around one structure. As Captain Crawford had predicted, 1st Platoon was clearly engaged with a significant enemy command and control point. To the Soldiers on the roof, the enemy appeared determined to hold the position at all cost. The willingness of the Taliban to stand and fight however, was a benefit to the Americans and their ANA counterparts. From his new strongpoint,

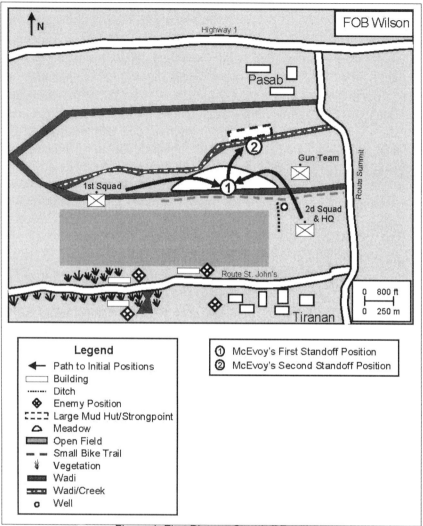

Figure 4. First Platoon Standoff Positions.

McEvoy had achieved an excellent standoff distance from the enemy while still maintaining the ability to observe their movements. While the Taliban appeared resolute; McEvoy and his platoon were equally unwavering in their determination to kill as many of the enemy as possible.[47]

Within minutes, Drake brought down a curtain of steel on the enemy targets. As Ingram observed the fire through his scope, he relayed the battle damage assessment (BDA) to Drake and McEvoy. Ingram also helped Drake adjust his fire. Taliban casualties began to mount as the two Soldiers continued their team effort and the howitzer rounds slammed into their targets. While the fight continued, Drake also directed rotary-wing firing, alternating these tasks between the indirect-fire missions. A US Army officer would report after the fight that Drake was "an invaluable asset with his calm and effective actions that brought an immediate and deadly result to the enemy." In fact, ISR sources had already picked up enemy communications in which a Taliban fighter told his comrades, "It's raining really bad, we've got to go."[48]

As the artillery barrage intensified, Ingram managed to hit several enemy fighters with shots from his M-14. The gun teams on the roof also kept up a steady stream of fire against the Taliban fighting positions. "We were firing at them as best we can…just kind of walking the rounds down as best we can," Doty said after the engagement. While 1st Platoon continued to fire, enemy rounds occasionally whizzed past their heads.[49]

Peering into the scope of his M-14, Ingram could plainly see the artillery impacting on the enemy positions. As he continued to scan the sector, he was surprised to see a Taliban open top truck with approximately 12 fighters in the back drive through the smoke and debris. Careening down a small dirt path, the truck abruptly stopped next to one of the enemy occupied buildings. The young marksman continued to watch in amazement as the Taliban fighters, armed with rocket propelled grenades (RPGs), heavy machine guns, and AK-47s jumped down from the truck and scattered in different directions. Ingram managed to hit one of the fighters as he ran for cover. He directed the M240G gun teams onto the other scurrying targets.[50]

The truck promptly sped away. As the dust cleared, Ingram spotted one of the enemy fighters brandishing a white Taliban flag. The man "started waving it at us, you know, taunting us," Ingram said. McEvoy thought the fighter was "just flipping us off, like, 'all right, come and get it.' " Ingram fired at the man but couldn't hit him. When Drake attempted to adjust fires on to the area, he received no response. Irritated, Drake asked, "Hey what happened to those rounds?" To their great consternation, McEvoy and Drake realized that the Squadron TOC was shifting its fires to other targets in the area. "I remember there being a lot of contention between me and the TOC on when and where to fire," McEvoy said. "From my viewpoint, I was shooting at the biggest threat to myself and my guys. Every once in a while, I'd jump on fires [network] to voice my concern over how targets were being prioritized."[51]

The TOC was indeed directing fire on other targets. Crawford would later write that, "The focus in the TOC was on moving elements coming to and from the fight. We utilized the ISR platforms to then call for fire with Apaches and Kiowas serving as forward observers. We also used the UAVs as observers for several fire missions." After the fight, McEvoy fully understood what the TOC was trying to do. "I later found out they were shooting at reinforcements, squirters, and other targets they could see with ISR. They also had a better view of the whole battlefield and the town of Tiranan," McEvoy wrote afterward. In the end, McEvoy came to understand the process but during this engagement he grew especially irritated over his inability to direct fire at what he considered the biggest threats to his platoon.[52]

Through the use of numerous ISR assets, the squadron TOC killed and wounded many enemy fighters that McEvoy and his Platoon could not see. As Taliban reinforcements rushed forward to protect their command post, their movements and signals allowed the TOC to capture prodigious amounts of intelligence. In fact, the longer 1st Platoon remained in contact, the more intelligence Crawford captured through his ISR sources. "What they wanted us to do," McEvoy said, " was go out there and just hold the enemy, engage the enemy…long enough so that they could develop the situation, and then they could call higher assets onto it while we're… fighting [them]. We held [them] for three hours there and they developed a lot of information about Tiranan."[53]

First Platoon was exhausted after three to five hours of intense combat. Ammunition and water were also running dangerously low. As he gazed at his tired and mud spattered Soldiers scattered around the hut, McEvoy decided to head back to FOB Wilson. "After a three-hour firefight, your adrenaline's just completely flat lined, you're out of water, out of energy," he said. "The dudes were just burnt," McKinnis recalled. Pulling out his map, McEvoy promptly picked out a route back to FOB Wilson as the men gathered their gear and prepared to move. Although the unit pulled away from the hut as if they were breaking contact, fire from the enemy had long since ceased, testimony perhaps to the devastating effects of the artillery and Kiowas.[54]

The platoon and the ANA moved north in a bounding overwatch formation. As they slogged back toward FOB Wilson, the NCOs maintained discipline, keeping everyone sharp. "They did a really good job," McEvoy said. Undoubtedly, many of the Soldiers thought about Chandler and speculated about his condition. After crossing north of Highway 1, the platoon turned east. Hattelsohl recalled that the north side of the highway was "known to be a lot safer."[55]

The Soldiers of 1st Platoon and their ANA complement arrived back at FOB Wilson without incident. As they walked through the gate they were greeted by the squadron command sergeant major and their first sergeant, who informed 1st Platoon that Chandler would be fine. As an added bonus, the 1st Platoon soon learned that intelligence sources had picked up enemy communications which indicated the "Taliban were surprised that the new unit showed teeth." It was the first of many disrupt-and-destroy missions the 1st Platoon would perform during its yearlong tour and it did, in fact, set the tenor for future operations.[56]

Aftermath

The combat at Tiranan proved the largest and longest fight of 1st Platoon's entire tour. The precise number of enemy casualties proved impossible to calculate. However, intelligence would later indicate that at least three Taliban fighters were killed and approximately five were wounded. Exact numbers are hard to determine. Even though it was their first encounter with the enemy, McEvoy's Soldiers performed courageously and skillfully. The platoon leader would later assert that although this was his first firefight, all of his NCOs were veterans of numerous combat tours. From the first enemy shots, McEvoy was determined to allow his NCOs to develop the fight. "Every squad leader did what they had to do for… five minutes, whatever it took to get the advantage, [and] report up to me," he stated. McEvoy was convinced that a Type A leader trained to make every decision "does not translate to a good infantry platoon, or a good maneuver platoon of any kind. I was a platoon leader for over a year and a half before I even went to Afghanistan," he pointed out. "I had worked with these guys, these guys respected me but you have to let them do their job. You need to know what they're doing, you need to know what they're supposed to do but don't do their job for them as it's [going to] end up having poor results or disastrous results for the platoon and Soldiers."[57]

The fight at Tiranan also reinforced several leadership principles for McEvoy's NCOs. "Just go slow, deliberate, do everything deliberately and slowly," McKinnis pointed out. Faggiano believed it was all about the basics and training. "It's all about basic skills…basic Soldier discipline and basic soldiering skills." Hattesohl would later state, "Never take the same route twice…Never take an easy route…If you see a trail, if you have to cross it, clear it, cross it, don't walk on it. It's too easy."[58]

The tactical gains accomplished by McEvoy's patrol were truly impressive. "The major result of this action was our ability to show the Taliban we would stand and fight rather than withdraw back to the COPs at the first sound of gunfire," wrote the Squadron S2, Matthew Crawford.

"McEvoy and his Platoon conducted a sustained five-hour firefight and never showed any sign of tactical impatience or concern. This shook the courage and morale of the local Taliban fighters who were used to dominating the battlefield by pushing patrols back to the safety of the COPs...The key result of this five-hour firefight was that we kicked the Taliban's ass in their own backyard." Without a doubt, the fight at Tiranan established a successful formula for 1-75 CAV and helped lay the groundwork for larger operations. The action also solidified the bond between the Soldiers of 1st Platoon and demonstrated their skilled, unwavering dedication in eradicating the Taliban in the Zhari District.[59]

For their actions, Mckinnis, Faggiano, Fraker, Hattesohl, Drake, and Chandler were awarded The Army Commendation Medal for Valor. For the duration of their tour in the Zhari District, 1st Platoon would continue conducting disrupt-and-destroy missions against the insurgents and their infrastructure. Many of 1st Platoon's "special missions" remain classified. The HHT commander however, would later suggest that McEvoy's platoon "conducted more missions than the rest of the squadron combined and more missions than any other element in RC South." Without doubt, 1st Platoon contributed significantly to eradicating the Taliban in the Zhari District.[60]

Notes

1. Carl Forsberg, "Counterinsurgency in Kandahar: Evaluating the 2010 Hamkari Campaign" Institute for the Study of War, December 2010, *Afghanistan Report*, 6, 23; Captain Matthew A. Crawford, e-mail to Matt M. Matthews, Combat Studies Institute, Fort Leavenworth, KS, 30 June 2011; Captain Riley E. McEvoy, e-mail to Matt M. Matthews, Combat Studies Institute, Fort Leavenworth, KS, 14 July 2011.

2. Forsberg, 7, 12, 14.

3. Nick Schifrin and Matt McGarry, "Battle for Kandahar, Heart of Afghanistan's Taliban Country," ABC news, 25 May 2010, 1, http://abcnews. go.com/International/Afghanistan/battle-kandahar-heart-taliban-country/ story?id=10729732 (accessed 6 April 2011); Emma Graham-Harrison, "Why the Military Plays Down Vital Afghan Battle," *Reuters*, 31 October 2010, http://www.reuters.com/article/2010/10/31/us-afghanistan-kandahar-idUSTRE69U0EI20101031 (accessed 11 April 2011).

4. McEvoy, e-mail, 14 July 2011.

5. McEvoy, e-mail, 14 July 2011; Crawford, e-mail, 30 June 2011.

6. Captain Riley E. McEvoy, interview by Matt M. Matthews, Combat Studies Institute, Fort Leavenworth, KS, 22 June 2011, 30-31; McEvoy, e-mail, 5 August 2011.

7. Captain Matthew Crawford, interview by Matt M. Matthews, Combat Studies Institute, Fort Leavenworth, KS, 20 April 2011, 10-11.

8. Crawford, interview, 15, 17.

9. Crawford, interview 17.

10. Sergeant Jesse Hattesohl, interview by Matt M. Matthews, Combat Studies Institute, Fort Leavenworth, KS, 23 June 2011, 2.

11. Sergeant First Class Derek Leach, e-mail to Matt M. Matthews, Combat Studies Institute, Fort Leavenworth, KS, 11 July 2001; Leach, e-mail, 5 July 20011; Hattesohl, interview, 3; Leach, e-mail, 6 July 2011.

12. McEvoy, e-mail, 11 July 2011; Leach, e-mail, 11 July 2011.

13. Leach, e-mail, 11 July 2011.

14. Leach, e-mail, 5 July 2011; McEvoy, e-mail, 11 July 20011.

15. Hattesohl, interview, 3-4; Leach, e-mail, 5 July 2011.

16. Hattesohl, interview, 4.

17. Hattesohl, interview, 4; Staff Sergeant Victor Faggiano, interview by Matt M. Matthews, Combat Studies Institute, Fort Leavenworth, KS, 22 June 2011, 11-12. McEvoy, e-mail, 12 July 2011.

18. Faggiano, interview, 22 June 2011, 10, 13-14; McEvoy, e-mail, 12 July 2011; Staff Sergeant Timothy McKinnis, email to Matt M. Matthews, Combat Studies Institute, Fort Leavenworth, KS, 12 July 2011.

19. McKinnis, e-mail, 12 July 2011; McEvoy, e-mail, 12 July 2011.

20. McKinnis, e-mail, 12 July 2011; McEvoy, e-mail, 12 July 2011; Hattesohl, interview, 23 June 2011, 5, Faggiano, interview, 14.

21. Hattesohl, interview, 6; Sergeant Zachary Fraker, interview by Matt M. Matthews, Combat Studies Institute, Fort Leavenworth, KS, 22 June 2011 4; Specialist Corey Doty, interview by Matt M. Matthews, Combat Studies Institute, Fort Leavenworth, KS, 22 June 2011, 3; Specialist Robert Kruse, interview by Matt M. Matthews, Combat Studies Institute, Fort Leavenworth, KS, 22 June 2011, 3; Specialist Andrew Ingram, interview by Matt M. Matthews, Combat Studies Institute, Fort Leavenworth, KS, 22 June 2011, 4-5; McKinnis, e-mail to Matt M. Matthews, 12 July 2011.

22. Narrative Page, Narrative To Accompany The Award Of The Army Commendation Medal For Valor To Sergeant Zachary Fraker, no date; Narrative Page, Narrative To Accompany The Award Of The Army Commendation Medal For Valor To Sergeant Jesse Hattesohl, no date.

23. Faggiano, interview, 15.

24. McEvoy, e-mail, 13 July 2011; Faggiano, interview, 15-16.

25. Hattesohl, interview, 6; Faggiano, interview, 16.

26. Sergeant Joseph Lee, interview by Matt M. Matthews, Combat Studies Institute, Fort Leavenworth, Kansas 22 June 2011, 15; Chandler, interview, 4.

27. Narrative To Accompany The Award Of The Army Commendation Medal For Valor To Sergeant Zachary Fraker, no date; Narrative To Accompany The Award Of The Army Commendation Medal For Valor To Sergeant Jesse Hattesohl, no date.

28. McEvoy, interview, 17; Specialist Mark Drake, interview by Matt M. Matthews, Combat Studies Institute, Fort Leavenworth, Kansas 22 June 2011, 5.

29. Narrative to Accompany the Award Of The Army Commendation Medal For Valor To Sergeant Victor Faggiano; Faggiano, interview, 16-17, 20.

30. Faggiano, interview, 20.

31. Faggiano, interview, 17; Crawford, e-mail, 14 July 2011.

32. Faggiano, interview, 17-18; Chandler, interview, 5-6.

33. Faggiano, interview, 18-19.

34. Kruse, interview, 3; McEvoy, e-mail, 13 July 2011.

35. Faggiano, interview, 20-21; McEvoy, interview, 19.

36. McEvoy, interview, 19.

37. Lee, interview, 18-19; Chandler, interview, 7-8.

38. McEvoy, interview, 21; Hattesohl, interview, 8-9.

39. Hattesohl, interview, 10-11; McKinnis, interview, 2.

40. Hattesohl, interview, 11-12; Fraker, interview, 11; McKinnis, interview, 2.

41. McKinnis, interview, 2-3; Narrative To Accompany The Award Of The Army Commendation Medal For Valor To Staff Sergeant Timothy McKinnis, no date.

42. McKinnis, interview, 3-4.

43. Hattesohl, interview, 14-15.

44. Faggiano, interview, 22-23.

45. Faggiano, interview, 23.

46. Faggiano, interview, 23; Hattesohl, interview, 15; McEvoy, e-mail, 11 July 20011 and 12 July 2011; Crawford, e-mail, 28 July 2011; Ingram, interview, 6.

47. Crawford, e-mail, 19 July 2011; Hattesohl, interview, 24.

48. Drake, interview, 5; Ingram, interview, 6; Narrative Page, Narrative to Accompany the Award Of The Army Commendation Medal For Valor To Private First Class Mark Drake, no date; McKinnis, interview, 5.

49. McEvoy, interview, 22; Doty, interview, 5.

50. Ingram, interview, 7; McEvoy, interview, 23.

51. Ingram, interview, 8; Drake, interview, 6; McEvoy, e-mail, 14 July 2011.

52. Crawford, e-mail, 14 July 2011; McEvoy, e-mail, 14 July 2011.

53. McEvoy, interview, 23-24; Crawford, e-mail, 14 July 2011.

54. McEvoy, interview, 25-26; McKinnis, interview, 5.

55. McEvoy, interview, 26; Hattesohl, interview, 19.

56. McEvoy, interview,26; Hattesohl, interview, 19.

57. McEvoy, interview, 32.

58. McKinnis, 6; Faggiano, interview, 27; Hattesohl, interview, 21.

59. Crawford, e-mail, 31 August 2011.

60. McEvoy, e-mail, 2 August 2011; Crawford, e-mail, 30 June 2011.

Trapping the Taliban at OP Dusty
A Scout Platoon In Zhari District
by
Kevin M. Hymel

The Black Hawk helicopter kicked up huge clouds of dust as it slowed to approach Observation Post (OP) Dusty. The men below, members of a scout platoon that had fought all day in an isolated mud-walled compound surrounded by insurgents, desperately needed ammunition. Soldiers on the compound's roof provided suppressing fire but the enemy still managed to target the hovering chopper, forcing its crew to grab rifles and return fire. The Soldiers had laid out markers for two drop spots. The Black Hawk just had to move a few feet over the western wall to drop the supplies. Instead, to the horror of the ground troops, the crew tossed the ammo packages outside the compound's walls. The scouts cut loose with a long string of curses, knowing they would have to leave the safety of the OP to replenish their supplies.[1]

The men in OP Dusty were there on a special mission to occupy an isolated compound before sunrise and fight the enemy until relieved by friendly forces. They were the linchpin to Operation NASHVILLE I, an attack by the 2d Battalion, 502d Infantry Regiment (2-502 IN), where they acted as the blocking force for the battalion.

Background

In 2010, Southern Afghanistan's Zhari District, 33 kilometers west of Kandahar City, was an insurgent hotbed filled with munitions caches and bomb making factories, all defended by a complex system of reinforcing defensive positions and belts of improvised explosive devices (IEDs). Taliban insurgents repeatedly attacked International Security Assistance Forces (ISAF) vehicles traveling Highway 1, the main east-west route between Gereshk and Kandahar City. The Americans understood Zhari's importance. "The Taliban in Pakistan have said that if they lose Zhari they'll be forced out of Kandahar," Lieutenant Colonel Peter Benchoff, the 2-502 IN battalion commander told his men. "So goes Zhari, so goes Kandahar City."[2]

American surge forces began arriving in southern Afghanistan in the end of 2009. By September 2010, the 2d Brigade Combat Team (2 BCT), 101st Airborne Division (Air Assault) assembled in Zhari north of Highway

1. The brigade consisted of the 1st Battalion, 502d Infantry Regiment (1-502 IN), the 2d Battalion, 502d Infantry Regiment (2-502 IN) and the 1st Squadron, 75th Cavalry Regiment (1-75 CAV). The Soldiers of 2-502 IN were surprised at what they found around Forward Operating Base (FOB) Howz-e Madad north of Highway 1. "The enemy was completely unfazed by our presence," said Lieutenant Colonel Benchoff. "They were primarily targeting private security convoys that moved up and down the highway." The enemy controlling the area had been threatening the locals with bodily harm, "chopping off heads and ears" Benchoff said. Captain William Faucher, the 2-502d's scout platoon leader put it more concisely, "It literally was the wild west. No civilians resided in the buildings south of Highway 1. Every house was a potential enemy bunker."[3]

Unlike the mountainous regions of eastern Afghanistan, Zhari's topography was flat. Grape, marijuana, and poppy fields covered the terrain with *wadis* (dry river beds) cutting through the landscape. The Taliban used the terrain to build defensive positions and belts of IEDs. Americans referred to the district's dense vegetation as "the Green Zone."[4]

The Plan

Operation NASHVILLE I was part of Operation DRAGON STRIKE, the brigade's effort to drive the enemy away from Highway 1. In a series of consecutive, mutually supporting attacks, the goal of DRAGON STRIKE was to force the enemy as far south as the Arghandab River which is 13 kilometers south of Highway 1. Benchoff designed NASHVILLE I, scheduled for 26 September 2010, as a simultaneous attack from multiple directions against Taliban occupied areas south of Highway 1. He planned to fly one of his companies over the enemy's IED belts and other defenses while two other companies assaulted the enemy's flanks.[5]

Bravo Company would air-assault one kilometer directly south from FOB Howz-e Madad and attack the town of Baluchan, the main objective. Then two platoons of Alpha Company would attack dismounted from the town of Paultercon, to Bravo Company's west while Delta Company attacked with armored vehicles and heavy equipment from Spin Pir to Bravo's east. The three companies would conduct their operations with partner Afghan National Army (ANA) companies. Delta Company would also cut a road as it advanced with the help of a British Engineer company. The six companies would converge around Baluchan where Alpha and Bravo, with their Afghan allies, would "Back Clear"—turning north and systematically clearing the area of the Taliban, IED, and weapons manufacturing—while Delta and its ANA counterpart company continued to drive south. "We wanted to seize the ground so we could hold it and see what we could trap," explained Benchoff.

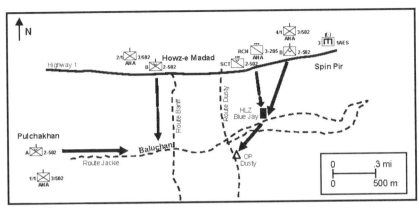

Figure 1. Operation NASHVILLE I, 26 September 2010.

Bravo Company would be isolated in Baluchan while the other companies fought from the east and west toward it. To assist Bravo, Benchoff decided to air-assault a platoon 1,500 meters east/southeast of Baluchan to interdict Taliban reinforcements that might move toward Bravo Company. In addition, the platoon would prevent the enemy from flanking Delta Company. As Delta pushed south, Benchoff hoped the platoon would trap the enemy between itself and Delta, a classic hammer-anvil maneuver or what Benchoff called "a mini kill sack."[6]

Benchoff picked a compound for his platoon and named it OP Dusty after Route DUSTY, the north-south road next to it. If the enemy decided to attack or retreat from Bravo Company's position, Route DUSTY was the obvious path and OP Dusty was the ideal blocking position. Benchoff chose to insert his scout platoon near the OP and occupy the compound before sunrise, hoping to trick the Taliban into thinking the route was open. The scouts would land at helicopter landing zone (HLZ) BLUE JAY and maneuver to the OP. Positioned on the eastern edge of the maneuver area more than 500 meters south of Highway 1, the scout platoon would be isolated, much like Bravo Company. "It's kind of like putting in your paratroopers in Normandy," Benchoff said. "You secure the outer perimeter and then you can take your time getting off the beach."[7]

Scout platoons normally conduct reconnaissance and security missions for infantry battalions but Benchoff wanted his scouts to do more. He had trained them in ambushes, sniper work, and operating the M240B medium machine gun. Benchoff had also used his scouts before in similar roles. "Don't limit yourself by what doctrine says," explained Benchoff about his approach to tactical problems. For NASHVILLE I, he would have continuous daylight air coverage for the operation so he attached the battalion's Joint Terminal Attack Controller (JTAC), Air Force Senior Airman Nathan Archambault, and the battalion Joint Forward Observer (JFO), Staff Sergeant Jason Thompson[8] to the platoon.

The scout platoon consisted of three reconnaissance teams of five or six Soldiers and a sniper section of nine men divided into teams of three including a sniper, a spotter, and a security man. A seven-man reconnaissance platoon of ANA Soldiers, entering combat for the first time, would accompany the unit. The platoon leader, Captain William Faucher, had served with the battalion for three years and had deployed to Iraq in 2007. His platoon sergeant, Sergeant First Class Justin Bosse, had been on four combat tours including Afghanistan in 2002. Staff Sergeant James Spear led the 1st Recon Squad and Sergeant Eric Ammerman, a former Marine, led the 2d Recon Squad. The two men complemented each other. "I'm just kind of carefree," Ammerman said, "Spear is very meticulous about every detail but when you put the two of us together, we work extremely well." Ammerman had served three tours in Iraq, while Spear had served two, both with the 2-502 IN. Staff Sergeant Chris Spoerle, the 3d Recon Squad leader, also had two tours in Iraq. Faucher described him as a "quiet guy, somewhat reserved but all action when the time came." Staff Sergeant Sterling Stauch, whom Faucher described as a blond-haired, blue-eyed "surfer boy," led the sniper squad. He had also served two tours of Iraq.[9]

Each scout packed 32 20-ounce bottles of water, eight MREs (Meals Ready to Eat), and 20 empty sandbags. Most were armed with M4 rifles with attached M203 grenade launchers. Some had M14 Enhanced Battle Rifles and three men were equipped with M249 Squad Automatic Weapon (SAW) machine guns. Squad leaders used M320 grenade launchers while the snipers carried either the M107 Barrett or M24 sniper rifle. To provide extra firepower, Faucher had two of the sniper teams each carry an M240B medium machine gun with tripod and ammunition. He also added four AT4 anti-armor weapons to the platoon's arsenal. Each man also had between 240 and 270 rounds of rifle ammunition. "We had a lot of extra ammunition," said Faucher, "because I had a feeling it was going to be a fight."[10]

Air Assaulting to OP Dusty

Three hours before sunrise on 26 September, a US Army CH-47 Chinook helicopter touched down in HLZ BLUE JAY. The rear ramp lowered and the scouts hustled out, each man slinging a 100-pound rucksack on his back. The first men out formed a defensive perimeter. Staff Sergeant Thompson, the JFO, twisted his ankle as he stepped off the ramp. "Man, I'm sorry," he told Sergeant Eric Ammerman, "I'm sorry, I'm sorry." He knew the ankle was bad but he also knew he was the platoon's only trained JFO. His skills would be needed for the coming operation. He decided to, "Suck it up and do this."[11]

Once the helicopter took off, Sergeant Damon Sawyer took point. Knowing that the most direct or easiest path would be mined, Sawyer led the scouts over the worst terrain he could find: across an eight foot canal, through thick grape fields, and over various mud walls. Men fell down, muddied their weapons, but got up and struggled on. The entire journey took about three and a half hours.[12]

Thompson carried his own rucksack despite his painful ankle. He twisted it four more times during the journey. "You just keep moving," Thompson recalled, "you keep your body going forward." He dropped his rucksack only once during the difficult climb out of the canal. "It's almost like spinning your wheels and getting your truck stuck." Captain Faucher knew that if Thompson could not walk, four scouts, already carrying 100 pound rucksacks, would have to carry him on a liter while another Soldier would have to lug two rucksacks, "which would have been a level of pain that I could hardly imagine," said Faucher.[13]

The chosen compound stood in the middle of grape fields overlooking Route DUSTY, the expected enemy avenue of movement. The scouts' new OP was approximately 25 square meters, surrounded by a four meter high mud brick wall with an entrance in the southeast corner. A line of four two-story buildings made up the northern wall, tall enough to give gunners a 360 degree view of the surrounding terrain. The roofs were arched, giving the look of numerous bumps and making for excellent fighting positions. The center of the compound contained a large courtyard with a storage hut on the southern wall. Unfortunately, a two story grape hut stood just a few meters north of the compound blocking the platoon's fields of fire in that direction. Around the OP trees, grape huts, compounds, and mud walls dotted the terrain.[14]

The platoon arrived at the compound before sunrise. Sergeant First Class Bosse led three men through the entrance and, using flashlights, made sure all the rooms were unoccupied. They then tossed blocks of C-4 demolition charges into every room to neutralize any IEDs. Twelve explosions followed. Two of the explosions caved in two roofs along the north wall. Charges were also detonated in the grape hut, caving in the eastern part of its roof. Once the dust settled, the team made a secondary sweep and then the entire platoon entered into the compound.[15]

Once in, the scouts dropped their rucksacks. The 1st Recon Squad used the collapsed buildings to climb to the roof on the northern wall and prepare defenses. Below, most of the scouts began filling sandbags while others dragged a tractor plow from the compound to block the entrance. The platoon medic, Sergeant Jason McMillan, treated Thompson's ankle.[16]

The Attacks Begin

About half an hour later, as the sun cracked the horizon, a single shot rang out. The scouts on the roof tried to locate the shooter while the men in the compound worked to fortify their positions. A few minutes later, a second round cracked overhead. About 15 minutes after that, the enemy opened fire from the south with AK-47 assault rifles. Then every compound and hut surrounding OP Dusty seemed to come alive with fire. The scouts had been taught that in combat there were two fights, a 15-second fight and the actual fight. If they could dominate the first 15 seconds, they would set the tone for the rest of the engagement. The scouts knew they had to dominate the enemy, beat them back, and prevent them from moving against Bravo Company.[17]

Sergeant Ammerman and Staff Sergeant Spear were on the rooftop scanning for targets when the enemy fire intensified. Ammerman called for someone to throw the SAW up to him. Once he got it, he stood and sprayed the area suppressing the enemy fire. Specialist Jason Savant climbed onto the roof with his M240B medium machine gun and pulled the trigger. Two rounds fired before the weapon jammed. During the infiltration, it had become fouled with dirt and brush. Spear and Savant quickly opened the weapon, lubricated it with oil, and soon had it firing.[18]

The scouts fired their M203s and M320 grenade launchers at insurgent locations. As the firing intensified, Captain Faucher ordered more men to the roof. Everyone wanted to be part of the firefight but Staff Sergeant Stauch knew there was plenty of work elsewhere. He ordered Specialist McMillan to gather anyone he could to fill sand bags. The completed bags were then passed to the men sprawled on the roof. For the rest of the day Soldiers fought while simultaneously building up their positions. Spear made sure everyone was staying down. "I did not want anyone to get their head [shot]," he later said. "I was a stickler."[19]

When the enemy opened fire from a grape hut to the east, Specialist Savant ran across the roof and began firing with his M240B. "You could actually see [the insurgents] moving in the grape hut," he recalled but his weapon could not penetrate the hut's thick walls. An ANA Soldier nicknamed "The Russian", also opened fire. "He started spraying to the east right off the bat," said Staff Sergeant Stauch. "[He] didn't care. [He didn't have] doubt or fear or anything."[20]

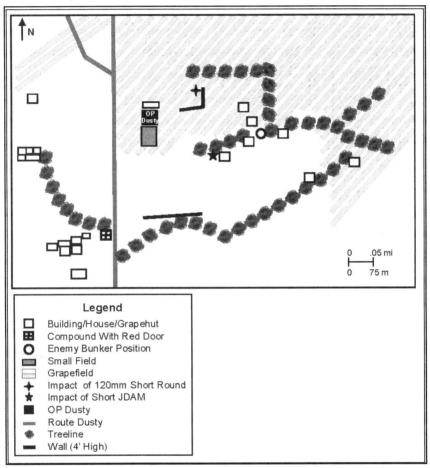

Figure 2. Scout Platoon Defends OP Dusty.

Thompson radioed the mortar men back at FOB Howz-e Madad and gave them target coordinates but enemy fire pinned down both him and Faucher before they could detect where the rounds had landed. Neither he nor Thompson could "physically move onto a knee and poke our heads over to observe fires," said Faucher. Instead, they relied on Ammerman and Savant who were covering each other. While Savant laid down fire with his M240B, Ammerman stood up, fired his SAW, and watched the mortar rounds explode. Ammerman then shouted their location to Thompson who adjusted for the next round. As the rounds came in, Thompson yelled

"Splash!" warning Ammerman and Savant to duck. After each explosion, Ammerman and Savant repeated the process.

Thompson called for mortar rounds to impact within 200 meters of his location, well within the defined Danger Close distance for 120mm mortars. One of the rounds fell a little too close at 70 meters. "Hey!" Thompson shouted into his radio, "the last round was pretty short, just check yourselves again." No other rounds came in short. The first six rounds used delayed fuses, penetrating the ground before exploding. The next rounds used proximity fuses for air bursts. Finally, the 20th round impacted right on the enemy. The explosion propelled an arm and a torso into the air. "It split him in half," said Ammerman. The men cheered. After 15 minutes, the initial attack died down.[21]

Once the engagement concluded, Faucher organized the platoon's squads into shifts with the 2d Squad on the roof from 0700 until 1100 followed by the 1st Squad from 1100 until 1500 and the 3d Squad from 1500 until 1900. The rotation cycle allowed one squad to pull security on the roof for four hours, another to fill sandbags or pass ammunition to the Soldiers on the roof, while the third rested. Sandbag duty preceded roof security, ensuring the men would be fully awake by the time they faced the enemy.[22]

The enemy's attack had surprised the platoon's veterans. Normally, the insurgents waited to determine what the Americans were doing before attacking. This time they started shooting once they saw helmets on the compound's roof. Soon the enemy attacked again. The attacks blurred together. There would be eight that day but the scouts noticed certain patterns. The enemy began firing from the south drawing the platoon's fire, then fired from the east, west, and eventually north, using suppressing fire while closing on the OP. Major attacks lasted about 30 minutes and increased in intensity and when the enemy disengaged, they waited approximately 40 minutes before attacking again. The scouts used these lulls to prepare for the next attack.[23]

Heavy vegetation, trees, and canals provided ample cover and concealment for the insurgents. Compounds, grape huts, and walls also allowed them to pop up, fire a few rounds, and disappear. The scouts could only spot the enemy briefly as they ran along the tree lines making the snipers' job difficult. "It was impossible to get a lock on with a sniper rifle and take a clean shot with all that vegetation," recalled Staff Sergeant Stauch. The scouts also had trouble spotting tracer fire and instead relied on the smoke from the enemy weapon's discharge. They called the smoke "tossed salad" and shouted it out to pinpoint a shooter's location.[24]

The Soldiers on the roof continuously exchanged fire with the enemy. "We had to expend a lot of ammo," Specialist Marvin Speckhaus said. "We didn't see where they were coming from." Some of the scouts worked in teams. On the east wall, Sergeant Ryan Spinelli and Specialist Erik Howes alternated firing M203 rounds while the other reloaded. Later at the compound entrance, Howes directed some ANA Soldiers to return fire. Sergeant Mike Brilla and Specialist Joseph Wilhelm took turns firing an M107 Barrett sniper rifle while the other spotted. Brilla used it to shoot holes in a wall concealing an insurgent gunman. The .50-caliber bullets penetrated the wall but Brilla had no idea if he scored any hits. He later spotted an insurgent holding a radio and alerted Wilhelm, but before Wilhelm could fire, the man disappeared. Even if the insurgent had presented a target, Wilhelm could not have shot him. "If he only has a radio, you can't pop him," said Brilla. The rules of engagement prevented the scouts from firing on anyone without a weapon. "They'll shoot at you," said Brilla, then "they drop their weapons and they just walk away."[25]

On the rooftop, scouts fired at the insurgents until they disengaged, only to pop up firing in another location. "It was really hard to pin them down which you almost couldn't do," said Specialist Savant. The air filled with enemy fire. Specialist Alex Hull was scanning his sector with his M14 when enemy rounds started whizzing by his head but he held steady, searching for the enemy until Staff Sergeant Stauch pulled him down.[26]

The Taliban also fired rocket-propelled grenades (RPGs) at the OP throughout the day. They initiated one of their attacks by firing an RPG from the southeast which zipped just over Staff Sergeant Thompson's head. "If I could have frozen time," said Thompson, "I could have just reached up and probably touched it." When Sergeant Brilla saw an RPG flying at him and Wilhelm, he told Wilhelm to duck. Wilhelm asked why as the RPG whooshed over his head. "That's why," Brilla explained. The round ricocheted off the corner of the grape hut but failed to explode. Another insurgent fired an RPG from the northeast, which skipped off the wall beneath a scout, leaving a rooster tail of smoke. "They must have been close enough that it didn't arm," said Sergeant Sawyer, "or maybe it was just a bad rocket."[27]

Captain Faucher spent most of the day on the roof directing machine-gun and sniper fire along with mortar, close air support from fixed-wing aircraft, and close combat attack (CCA) missions from Army helicopters. As the men shot smoke grenades at targets, he called in gun runs by OH-58 Kiowa and AH-64 Apache helicopters, Air Force A-10 Thunderbolt IIs, and Navy F/A-18 Hornets. Faucher worked well with Senior Airman

Archambault, Sergeant Thompson, and radio telephone operator (RTO) Specialist Aaron Jonas who helped Thompson call in the aircraft. "Thompson, I want fire over there," Faucher would call out, or "Jonas, tell the Apache I want a Hellfire missile on that building." When Faucher saw the enemy fire an RPG from a tree line, he ordered Archambault to call in the A-10s to strafe the area. The pilots spotted four insurgents moving west and strafed them twice with their 30-millimeter cannons. The enlisted men's skills allowed Faucher to focus on the fight and decide where to place fires.[28]

Thompson and Archambault spent the day close to Faucher. Neither had ever served in combat nor had they worked together but they coordinated all of the captain's fire. "Luckily, both of us were confident enough in our abilities and we were confident in each other's abilities," Thompson said. "I'm dropping mortar rounds [and] he's talking aircraft on the targets. I'm stopping one mission, he's coming in doing strafes or dropping bombs with A-10s, then I'm bringing the CCA." Jonas also helped out by updating pilots on the ground situation so they were ready when Archambault contacted them. Thompson spent the whole day in pain. His ankle made it difficult to move around the compound and roof. "No matter what I took or what I did," said Thompson, "it just hurt."[29]

When Sergeant First Class Bosse was not in the compound directing the improvement of defensive positions, he was on the roof coordinating fire, making ammunition counts, and helping anyone who needed it. He didn't hesitate to stand up on the roof to point out targets. At one point when he felt the air assets weren't responding quickly enough to requests he shouted into the radio, "Can I get some goddamn air down here? We're getting shot at!" Thompson standing nearby thought, "All right. That's the way to request it!"[30]

By late afternoon the enemy's tenacity began to impress the scouts. "This is ridiculous," thought Thompson. Insurgents kept on attacking despite the scouts' volume of fire and Afghanistan's 110 degree temperatures. Scouts were firing so rapidly that their weapons burned their hands through their gloves. At one point the enemy attacked from every direction and closed within 125 meters. "These people are getting close," Spear told Faucher. "You know this is getting bad." The men were confident in their fighting ability and their air assets but they were concerned about their dwindling ammunition supply as the enemy crept closer. "We were either going to go down fighting," recalled Sergeant Spinelli, "or we were going to send them off."[31]

Resupply

The platoon was dangerously low on 40-millimeter smoke grenades, vital to marking enemy targets and 40-millimeter high-explosive rounds. The M240B ammunition was down to 200 rounds. Bosse had radioed battalion at about 1000 for an aerial resupply but the battalion RTO relayed that with all the platoon's air support, resupply was unnecessary. Bosse continued to make requests and kept getting the same answer from the battalion RTO. By early afternoon, he was fed up and demanded to talk to the battalion leadership directly. He soon reached the battalion commander and told him, "We need this aerial resupply."[32]

Finally, around 1600, a Black Hawk helicopter flew toward OP Dusty. Ammunition was packed into crates and body bags (speedball packages) for easy handling. The helicopter slowed as it approached drawing enemy fire as the scouts on the roof attempted to suppress the insurgents' positions. Faucher had no radio contact with the helicopter as it hovered over the west section of the compound. "They were so close," he recalled "If I jumped, I could have probably made it into the Black Hawk." That's when the Black Hawk crew threw the ammo packages outside the OP's western wall and into a grape field.[33]

As the Black Hawk flew away, Bosse picked a retrieval team of six men and led them out of the compound while scouts on the roof laid down suppressing fire and launched smoke grenades to obscure their movement. The retrieval team exited the southeast entrance and worked its way along the northern wall until turning south to find the supplies. Some of the speedballs had split open, littering the grape field with ammunition. Bosse set up a perimeter while the rest of the team tossed ammunition over the wall. Men on the other side caught it, wiped off the mud and threw it up to the roof. The retrieval team also stuffed their pockets with magazines and rounds and carried what they could, either in crates or body bags, back into the compound. They man handled one torn body bag back to safety. "It took about three of us to drag the portion of it that was still together," said Sergeant McMillan. After collecting as much as they could for about 20 minutes, the men made it back safely into the compound.[34]

Sundown

Before sundown, teams of Soldiers left the compound again to set up Claymore mines in dead space and possible avenues of approach around the perimeter. The teams placed the mines roughly 100 meters from the compound walls. "Some were pretty close, but they needed to be," said Sergeant Sawyer. "As long as you got good cover, you could stay 15 meters behind them." Bosse and Spinelli were placing their Claymore when a shot rang past. They charged back into compound as a scout covered them with the SAW.[35]

As evening approached, Staff Sergeant Thompson began calling in smoke rounds in hopes of burning down some of the surrounding brush and keeping the enemy out of previously occupied areas. The effort proved fruitless. Nothing caught fire and the enemy reoccupied the area the next day. "They definitely did not want to give up that terrain," said Thompson, who no longer limped when he walked.[36]

The enemy broke off its final attack around 1800 as the sun sank in the west. The roof and compound were strewn with empty shells from the constant fighting and sandbags were shredded from enemy fire. The men were covered in sweat and dust. The adrenaline rush that had sustained them through the last 12 hours of combat finally wore off and they found themselves exhausted. "I was like comatose," said Staff Sergeant Spear.[37]

None of the scouts had experienced such intense fighting in previous combat deployments. Sergeant Brilla thought the day's combat was more severe than any he had endured in Iraq. Two-tour veteran Staff Sergeant Stauch later said, "It was definitely the most contact I've ever taken." Thompson thought the fighting was worse than Iraq's Sunni Triangle, "Nothing that I have ever done will compare to [OP Dusty]." Specialist Eric Howes added, "I've never had so many bullets shot over my head in one day." Amazingly, despite eight enemy attacks with heavy volumes of fire and RPG rounds, none of the scouts had even been wounded.[38]

Captain Faucher needed a break from the non-stop action. He took off his helmet, sat down and decompressed for a little while. After eating an MRE, he huddled with Bosse to evaluate the enemy's patterns of attack. Then, while Faucher planned targets for the next day with Sergeant Thompson, Bosse took expenditure reports from the men. He wanted to minimize security and maximize the rest plan but after the intensity of the day's fighting, he decided to keep about 12 men on the roof at all times.[39]

During the night, the men replaced torn-up sandbags, distributed ammunition and cleaned their weapons. Most managed about four hours of sleep. Faucher received combat update briefs from battalion while some of the men listened in. Thompson spent the night calling in random illumination rounds, letting the enemy know that the scout platoon was awake and alert. Ammerman made the rounds all night, making sure the men on guard duty stayed on guard.[40]

The Second Day of Battle

As the sun rose the next morning, the enemy opened fire again. "It's like they were punching in a time clock at work," said Sergeant McMillan. Instinctively, the Soldiers began looking for targets asking, "Where's it coming from, where's it coming from?" Fire came from a compound with a red door to the southwest. The scouts laid down suppressing fire with an M240B machine gun as M-203s and sniper rifles added to the fray. "We let them have it," said Spear. The second fight for OP Dusty had begun.[41]

Hearing the commotion, Specialist Howes woke up, grabbed his rifle, and climbed up to the roof where he found Captain Faucher directing fire. By this time, the enemy was firing from various locations. "Hey, we're taking small arms fire from over here," shouted one of the scouts while pointing north. Rounds started flying over everyone's head. Spear fired his M320 at the location while Howes fired a number of red smoke grenades to mark targets for the helicopters that had been circling since dawn.[42]

Enemy fire intensified just as it had the day before. At the OP's entrance, Sergeant Brilla fired his M4 from behind the tractor plow until an RPG round flew in and bounced off the wall behind him, leaving a four-inch hole without exploding. "It really scared the hell out of me," he said.[43]

After two hours of action, the enemy's activity died down. The morning's first firefight was over. During the lull, Sergeant Brilla and Specialist Howes left the compound to set up additional Claymores. They placed two mines along the western wall before hurrying back inside without incident.[44]

The enemy attacked again 40 minutes later. Other attacks followed. Just like the previous day, the scouts lost track of the attacks as they waited on Delta Company's approach from the northeast. At one point, the fighting became so intense that Faucher did something platoon leaders were normally not supposed to, he fired his weapon. "I was told that if I was shooting, I was wrong." but when Sergeant Spear needed suppressing fire to shoot his M320, Faucher shouldered his M4 and fired. "I was kind of mad," he admitted, knowing that if he was shooting, the situation was dire. During one of the lulls, Spear returned the favor. He noticed Faucher standing, scanning for targets. Spear pushed him down. "I just saved your life," Spear told him. "No big deal."[45]

The men improvised when needed. With no sandbags to spare, Specialist Howes laid on the roof as Specialist Wilhelm slid his M110 sniper rifle over the small of Howes' back. The two men formed a "T" and timed their breathing to keep the rifle steady while Wilhelm shot at the elusive enemy. Elsewhere, Specialist Savant saw insurgents closing to about 25 meters from the compound walls. He fired down at them with his M240B. "By that time you could actually see them just sprinting away," Savant recalled. The enemy never pushed any closer to the OP.[46]

The compound with the red door was a constant nuisance. Sergeant Brilla thought a gunman there had specifically targeted him. "Whoever was in that compound had a pretty good bead on me," he recalled. When Brilla could get his head up, he fired M203 rounds at the compound without much effect. Captain Faucher called in repeated air strikes on the structure but the enemy continued to fire. "I think I put four or five 100-pound bombs in it," explained Faucher. "It would just not fall down. It was driving me nuts." He also had Archambault call in air strikes on a grape hut to the southeast. A-10s repeatedly strafed the building but, just like the compound with the red door, the enemy remained active.[47]

When insurgents started taking potshots at one of the Kiowa helicopters circling above, Staff Sergeant Stauch and Specialist Kenyon Jackson paid special attention to the tracers to determine their point of origin. After the fourth round, Stauch opened up with a SAW. He then fired a grenade followed by red smoke. The round landed short but bounced about 50 meters to the right spot. The Kiowa pilots spotted it and swooped down to engage two insurgents running through a tree line. Stauch was not sure if the enemy survived the gun run. "I know they were in really deep vegetation by the time the birds got eyes on." At least the enemy no longer fired on the Kiowa. [48]

At one point, the enemy fired an RPG that arced just above the Soldiers and impacted into the northern grape hut. Within seconds, Spear propped himself up, exposing himself, and prepared to fire an AT4 rocket launcher at the base of the smoke trail. Before pushing the trigger, he shouted "Backblast area clear!" Unfortunately, Captain Faucher did not hear Spear's call as he climbed up to the roof with Jonas, Thompson, and Archambault. All four men were caught in the backblast and temporarily lost their hearing. The blast also threatened to cave in a roof below where members of 1st Squad were resting. The men sprinted out of the room. Spear's shot silenced the shooter. No more RPGs were fired at OP Dusty.[49]

Responding to enemy fire from a tree line to the south, Senior Airman Archambault lined up a flight of US Navy F-18s armed with joint direct attack munitions (JDAMS). The first JDAM hit right on target but the second one fell within 125 meters of the compound. "It's incoming," shouted Archambault. "Get down!" Everyone dropped as the explosion rocked the compound. The blast almost threw Faucher off the roof. "It was like being hit with a brick in the chest," he explained. Shrapnel sprayed the compound and a piece smacked Specialist Howes in his rear, but did not penetrate.[50] The roof of the only remaining room caved in. The men resting there charged out, never to return.

As the smoke cleared, the scouts looked at Archambault. Some thought that he deliberately dropped the bomb short. "Are we about to get overrun and I just don't know it yet?" thought Staff Sergeant Stauch. "That was not me," Archambault immediately defended. That's all it took. The men recognized and trusted his abilities. Stauch called the pilot and, in his California accent, shouted, "Whoa! Whoa! Whoa! Like what are you doing?" His words brought a little humor to a tense situation. The Navy held up the next run to reconfirm their coordinates.[51]

By afternoon the scouts could hear and then see Delta Company approaching from the northeast. Delta Company employed M58 Mine Clearing Line Charges (MICLICs) to blast through IED fields and clear the enemy, sometimes firing the MICLICs over buildings to collapse them. "[The MICLICs] pretty much crush everything," said Sergeant First Class Bosse. As Delta drew near, Spear, concerned about friendly fire, warned the men on the roof, "Only fire 40-millimeter rounds. Don't exceed that." After that warning, if anyone used a rifle to engage the enemy, the men around him would shout, "Hey! Do not engage with the M4!"[52]

The Soldiers of Delta Company pressed close enough to trap some insurgents in a tree lined ditch between them and the northern section of OP Dusty. Instead of pressing their attack on the scouts, the enemy turned and directed their guns north. One of Delta Company's M-ATVs [MRAP] (Mine Resistant Ambush Protected All Terrain Vehicle) retaliated with blasts from a .50-caliber machine gun, sending rounds over scout heads and smacking into the compound walls. "Hey," shouted Sergeant Brilla, "that's a M-ATV out there. I can see them!" Some of the men thought the insurgents possessed a .50 cal. "What should we do?" someone shouted. Faucher yelled back, "Stay down dumb ass!" The scouts draped VF17 identification panels over the compound walls while Faucher quickly contacted Delta's commander, Captain Tim Price, and told him to immediately cease firing. The enemy in the ditch were finally suppressed by a combination of small arms fire and Apache helicopters firing Hellfire missiles and 30mm rounds.[53]

Figure 3. Senior Airman Nathan Archambault watches from the edge of the compound roof as a GBU-12, laser-guided bomb explodes on a grape hut outside of OP Dusty.

The compound with the red door had survived numerous attacks thus far but now Captain Price came on the radio net, asking Faucher, "Are you guys taking fire from that compound? It's going down." With that, an M1128 Mobile Gun System (MGS), a Stryker armored fighting vehicle with a 105mm cannon, sped up to the compound, braked to a dead stop, and started firing point blank. By the time it stopped, recalled Spear, "you could hear crickets." Soon after the MGS silenced the compound, British Trojans (armored engineer vehicles) rolled into the area bulldozing a road as they came. Once the Trojans reached the OP, they began using their hydraulic excavator arms to tear down the northern grape hut. While the Trojans provided the scouts a better field of fire, their endeavors brought an unexpected problem, "They stirred up a bunch of bees," said Sergeant Sawyer. "We asked them to stop."[54]

The scouts had spent the previous 48 hours battling a tenacious enemy but it wasn't until Delta Company showed up with its armored vehicles that the insurgents abandoned the battlefield. "We killed them with bombs and small arms," said Sawyer, "but they were not intimidated by us. They got very shook once those tanks [*sic*] rolled in." Sergeant First Class Bosse agreed with the importance of Delta's presence, "Once [Delta Company] got down there, all the fun was over." Around 1600 that afternoon,

Delta finally linked up with the scout platoon. Captain Price entered the compound and shook hands with Faucher. The two men then discussed the situation and made tentative plans for their next action.[55]

With Delta Company securing the area, it was time for the scouts to leave OP Dusty but with so many aircraft involved in combat around the district, no helicopters were available to take the men back to base. So, after two solid days of fighting, the scouts walked home. At least the kilometer march to the north would be on the dusty road of Route DUSTY instead of the terrain they had passed over two nights ago. Before they left, Bosse wryly joked that for two days the scout platoon had owned about 95 percent of the air assets in all of Afghanistan. Thompson slung his rucksack on his back. His ankle felt good enough for the trek out. As he walked out of the OP, Senior Airman Archambault turned to him and said, "The scout Platoon was not as fun as I thought it was going to be."[56]

Aftermath

The scout platoon had successfully prevented the enemy from attacking Bravo Company in strength and simultaneously lured the insurgents into a violent killing zone. Intelligence later reported that the two day fight had resulted in 15 to 20 insurgents dead. Enemy casualties might have been higher but a more definitive count was impossible because the Taliban, as usual, quickly removed their dead and wounded from the battlefield. No scouts had been wounded during the battle. They had served as the perfect blocking force for the battalion. "They were having the exact effect of what the design was," said Lieutenant Colonel Benchoff, "which was to get these guys [the enemy] to roll off them and roll into Delta Company. Then they got chewed up by Delta Company, which was entirely what was planned." Benchoff greatly appreciated what the scouts had done, stating, "That was probably the highest density of enemy attacks that we had [during the deployment]."[57]

NASHVILLE I was also a success. In one week, 2-502 IN had cleared Highway 1 and eliminated the enemy's ability to conduct an organized fight. "We had essentially achieved what was the brigade's primary objective for us for the entire operation," said Benchoff. For the rest of DRAGON STRIKE, the enemy never again fought as hard as they did at OP Dusty. The insurgents would only fire a few rounds at Soldiers or shoot an RPG and run. The intensity of battle never again matched those two days around OP Dusty.[58]

The scouts were understandably pleased with their performance. Faucher was particularly glad that the JFO, Staff Sergeant Thompson, despite his ankle injury, was able to call in Danger Close fire missions throughout the fight. The talents of both Thompson and the JTAC, Senior Airman Archambault, gave the platoon a critical advantage over the insurgents and helped hold the initiative during much of the battle. The four-tour veteran Bosse, told the men that with the exception of Operation ANACONDA, a brigade-sized operation back in 2002, this was the most intense combat he had experienced.[59] Thompson told the platoon RTO, Specialist Jonas, "I've never seen someone who's not a JFO handle aircraft the way that you did." Coming from an experienced forward observer, Jonas felt reassured.[60] When asked what they would have done differently, the scouts said only small things. Brilla and Spinelli would have packed more sandbags, Spear would have inserted right on top of the OP negating the tough trek and allowing the men to bring more ammunition, Thompson said he would not have twisted his ankle.[61]

Captain Faucher had clearly directed both the establishment of the position and the subsequent fight by placing teams around the OP, directing fire, and calling in support fires. He listened to suggestions from his noncommissioned officers and implemented them as he saw fit. The young officer's demeanor during the battle, and his placement of Soldiers on the roof impressed Staff Sergeant Bosse who explained that it was, "actually kind of rare for somebody who's only 23 or 24 to be that calm, cool, and collected through it all and make those good decisions." [62]

Through it all, the platoon worked like a well oiled machine. "Everybody knew their part, everybody had their part," said Bosse, "everybody stayed in their lane." Whether it was teaming up to cover a section of the roof, serving as a sniper rifle platform, or calling in continuous support fire, the scouts performed their mission with professionalism and esprit de corps. Faucher also credited simple fundamentals for the platoon's performance, "Every man did his job without fear or hesitation." [63]

Notes

1. Specialist Marvin Speckhaus, interview by Kevin Hymel, Combat Studies Institute, Fort Leavenworth, KS, 29 July 2011, 12; Sergeant Damon Sawyer, interview by Kevin Hymel, Combat Studies Institute, Fort Leavenworth, KS, 20 June 2011, 17; Staff Sergeant Jason Thompson, interview by Kevin Hymel, Combat Studies Institute, Fort Leavenworth, KS, 21 June 2011, 28; Captain Bill Faucher, interview by Kevin Hymel, Combat Studies Institute, Fort Leavenworth, KS, 15 June 2011, 33, 34; Staff Sergeant James Spear, interview by Kevin Hymel, Combat Studies Institute, Fort Leavenworth, KS, 17 June 2011, 22; Sergeant Eric Ammerman, interview by Kevin Hymel, Combat Studies Institute, Fort Leavenworth, KS, 20 June 2011, 26.

2. Anthony Loyd, "The Surge is Working—So Far," *Standpoint,* December 2010, http://standpointmag.co.uk/dispatches-december-10-the-surge-is-working-so-far-anthony-loyd-afghanistan-zhari, accessed May 14, 2011.

3. Lieutenant Colonel Peter Benchoff, interview by Kevin Hymel, Combat Studies Institute, Fort Leavenworth, KS, 16 June 2011, 10, 12; Faucher, interview, 3, 6.

4. Major Sean Brown, interview by Michael Doidge, Combat Studies Institute, Fort Leavenworth, KS, 10 June 2011, 14; Carl Forsberg, "Counterinsurgency in Kandahar: Evaluating the 2010 Hamkari Campaign" Institute for the Study of War, December 2010, *Afghanistan Report*, 6, 23.

5. Forsberg, 23.

6. Lieutenant Colonel Peter Benchoff, interviewed by Kevin Hymel, 23 August 2011, Combat Studies Institute, Fort Leavenworth, KS, 23 August 2011, 1.

7. Benchoff, 16 June 2011, interview, 6, 36-39; Faucher, interview, 13, 14, 17.

8. Benchoff, interview, 23 August 2011, 1; Faucher, interview, 18; Department of the Army, *Field Manual (FM) 3-21.20, The Infantry Battalion* (Headquarters, Department of the Army: Washington, DC, 2006) 1-8, 1-9.

9. Ammerman, interview, 46; Faucher, interview, 59; Captain William Faucher, e-mail to Kevin Hymel, Combat Studies Institute, Fort Leavenworth, KS, 16 August 2011.

10. Spear, interview, 5,6; Faucher, interview, 16, 29, 30, 37; Sergeant First Class Justin Bosse, interview by Kevin Hymel, Combat Studies Institute, Fort Leavenworth, KS, 15 June 2011, 7.

11. Bosse, interview, 2; Faucher, interview, 15, 19; Thompson, interview, 2, 5; Ammerman interview, 47.

12. Ammerman, interview, 4, 5; Sawyer, interview, 5; Faucher, interview, 9.

13. Thompson, interview, 6, 7; Faucher, e-mail.

14. Captain William Faucher, e-mail to Kevin Hymel, Combat Studies Institute, Fort Leavenworth, KS, 12 July 2011; Bosse, interview, 3; Thompson, 16.

15. Spear, interview, 7, 8; Thompson, interview, 9; Sergeant Ryan Spinelli, , interviewed by Kevin Hymel, Combat Studies Institute, Fort Leavenworth, KS, 28 June 2011, 5; Sawyer, interview, 16.

16. Jonas, interview, 17; Sawyer, interview, 22; Sergeant Jason McMillan, interviewed by Kevin Hymel, Combat Studies Institute, Fort Leavenworth, KS, 29 June 2011, 8.

17. Jonas, interview, 35; Ammerman, interview, 8; Faucher, interview, 22.

18. Specialist Jason Savant, interview by Kevin Hymel, Combat Studies Institute, Fort Leavenworth, KS, 23 June 2011, 9.

19. Specialist Eric Howes, interview by Kevin Hymel, Combat Studies Institute, Fort Leavenworth, KS, 20 June 2011, 7; Spear, interview, 18; Specialist Aaron Jonas, interviewed by Kevin Hymel, Combat Studies Institute, Fort Leavenworth, KS, 20 June 2011, 4; Staff Sergeant Sterling Stauch, interviewed by Kevin Hymel, Combat Studies Institute, Fort Leavenworth, KS, 29 June 2011, 12, 52; Ammerman, interview, 9, 10.

20. Savant, interview, 10, Stauch, interview, 36.

21. Savant, interview, 20; Thompson, interview, 13, 14, 15, 45; Ammerman, interview, 19, 20, 21, 40; Faucher, 56, 57.

22. Stauch, interview, 13, 32.

23. Faucher, interview, 29; Jonas, interview, 9; Ammerman, 18; Stauch, interview, 34.

24. Stauch, interview, 21, Sawyer, interview, 11; Spinelli, interview 24, 25; Sergeant Mike Brilla, interviewed by Kevin Hymel, Combat Studies Institute, Fort Leavenworth, KS, 29 June 2011, 11, 12.

25. Speckhaus, interview, 10; Howes, interview, 28; Brilla, interview, 12.

26. Savant, interview, 12; Stauch, interview, 16.

27. Thompson, interview, 21; Sawyer, interview, 21; Spear, interview, 14, 15, 24, Brilla, interview, 12, 13, 24, 36.

28. Speckhaus, interview, 6, 9; Thompson, interview, 21; Faucher, interview, 27, 30, 58; Jonas, interview, 10.

29. Thompson, interview, 18, 19; Staff Sergeant Jason Thompson, e-mail to Kevin Hymel, Combat Studies Institute, Fort Leavenworth, KS, 19 August 2011.

30. Thompson, interview, 60; Ammerman, interview, 48; Sawyer, interview, 41; Savant, interview, 40; Howes, interview, 33.

31. Thompson, interview, 20; Spear, interview, 19; Ammerman, interview, 52; Spinelli, interview, 13.

32. Bosse, interview, 14; Spear, interview, 19; Ammerman, interview, 27, 28, 49; Faucher, interview, 25, 30.

33. Speckhaus, interview, 12; Sawyer, interview, 17; Thompson, interview, 28; Faucher, interview, 33, 34; Spear, interview, 22; and Ammerman, interview, 26.

34. Spear, interview, 21, 22; Stauch, interview, 27, 30; Sawyer, interview, 19; Ammerman, interview, 26; McMillan, interview, 22.

35. Spinelli, interview, 10, 11; Sawyer, interview, 25, 26.

36. Thompson, interview 31, 32; Thompson, e-mail.

37. Faucher, interview, 41, 42; Bosse, interview, 23; Jonas, interview, 5; Sawyer, interview, 12; Spear, interview, 27.

38. Brilla, interview, 20; Stauch, interview, 35; Thompson, interview, 35; Howes, interview, 20.

39. Bosse, interview, 14; Ammerman, interview, 33; Faucher, interview, 46.

40. Spinelli, interview, 26; Faucher, interview, 46; Ammerman, interview, 33.

41. Thompson, interview, 39; McMillan, interview, 29; Spinelli, interview, 20; Spear, interview, 31.

42. Howes, interview, 22; Spear, interview, 16, 17.

43. Brilla, interview, 25; Sawyer, interview, 22.

44. Brilla, interview, 18.

45. Faucher, interview, 27, 28.

46. Savant, interview, 16; Howes, interview, 10, 23, 24.

47. Brilla, interview, 26; Howes, interview, 24, 25; Faucher, interview, 49.

48. Stauch, interview, 20, 32, 40.

49. Faucher, interview, 37, 61; Bosse, interview, 29; Thompson, interview, 42; Savant, interview, 28; Ammerman, interview, 22; Speckhaus, interview, 29.

50. Howes, interview, 25.

51. Faucher, interview, 52, 53; Jonas, interview, 26; Thompson, interview, 50, 51; Savant, interview, 26; Ammerman, interview, 41; Stauch, interview, 42, 43, 44.

52. Jonas, interview, 28; Bosse, interview, 34; Spear, interview, 35.

53. Jonas, interview, 23, 24; Faucher, interview, 54; Ammerman, interview, 37; Brilla, interview, 29, 30; Spear, interview, 36, Bosse, interview, 36.

54. Spear, interview, 37; Brilla, interview, 27; Sawyer, interview, 33, 35.

55. Sawyer, interview, 33; Bosse, interview, 49; Jonas, interview, 29.

56. Thompson, interview, 57; Savant interview, 42, 43; Brilla, interview, 23.

57. Faucher, interview, 68; Lieutenant Colonel Peter Benchoff, interview by Kevin Hymel, Combat Studies Institute, Fort Leavenworth, KS, 10 June 2011, 35.

58. Lieutenant Colonel Peter Benchoff, interview by Kevin Hymel, Combat Studies Institute, Fort Leavenworth, KS, 16 June 2011, 31; Thompson, interview, 56; Sawyer, interview, 30.

59. Thompson, interview, 34, Jonas, interview, 14, 32; Ammerman, interview, 34.

60. Faucher, interview, 44, 45; Jonas, interview, 32.

61. Howes, interview, 33; Spear, interview, 39; Jonas, interview, 32; Thompson, interview, 61; Faucher, e-mail.

62. Bosse, interview, 8.

63. Faucher, interview, 69.

Objective Lexington
Cougar Company under Fire in the Ganjgal Valley
by
Ryan D. Wadle, Ph.D.

In late March 2011, the men of Cougar Company, 2d Battalion, 327th Infantry Regiment (2-327 IN), also nicknamed Task Force *No Slack*, had nearly completed their year-long deployment to Kunar Province, Afghanistan. Over the past year, 2-327 IN conducted numerous operations throughout the province with several focusing on the rugged and remote regions of Kunar along the border with Pakistan. By the end of March, some Soldiers from the battalion had already returned to the United States and within the next three weeks, the remaining men would also leave Afghanistan. Still, the unit was ordered to launch a major offensive against insurgents concentrating in the Ganjgal Valley near the Pakistani Frontier. Some of the men of TF *No Slack* disliked the notion of starting a new operation so near to the end of their deployments but in spite of these reservations, they understood the necessity of continuing to pressure insurgent forces in their Area of Operations (AO). This final push against the enemy, Operation STRONG EAGLE III, would take the men of *No Slack* deeper into the remote and insurgent held Kunar Province than any American unit had operated before.

Background

The plan for Operation STRONG EAGLE III entailed nothing less than a direct assault on the home areas of insurgent leader Qari Zia Rahman and his subordinate, a man known only as Tamidullah. Since 2001, Coalition forces had only rarely entered the Ganjgal Valley. In 2009, insurgents had ambushed and killed four Marines and nine Afghan National Army (ANA) Soldiers at the head of the valley. Rahman, known by the abbreviation "QZR," used the area as a base of operations from which he dispatched fighters and weapons to other parts of Kunar Province. QZR had also established a clandestine radio station managed by Tamidullah through which he communicated with the local populace to discredit coalition activities. *No Slack*'s commander, Lieutenant Colonel Joel B. Vowell, hoped that neutralizing the insurgent headquarters in the Ganjgal would grant *No Slack*'s replacement, 2d Battalion, 35th Infantry Regiment (2-35 IN) - dubbed Task Force *Cacti* - time to fully establish itself in Kunar without facing the prospect of a major summer offensive from QZR's forces.[1] *No Slack* had engaged QZR's insurgents at several points during its deployment, most prominently during Operation STRONG EAGLE I in June 2010, but STRONG EAGLE III aimed to deal the deathblow to QZR's command group and establish coalition control over the Ganjgal Valley and the southern reaches of the nearby Ghakhi Valley, an area collectively codenamed Objective VIRGINIA.

To neutralize an isolated and hostile area such as Objective VIRGINIA required a complex operation involving all four of the battalion's maneuver companies. *No Slack* planners devised a deception operation to the north in the Sholtan Valley in order to draw fighters out of the key villages of Barawalo Kalay (Objective RICHMOND) in the southern Ghakhi Valley and Sarowbay (Objective LEXINGTON) at the eastern end of the Ganjgal. The battalion's Alpha Company would establish a blocking position to the north of the two villages to intercept and destroy any fighters who attempted to re-enter the main objectives. Meanwhile, one company would search and clear each of the villages, hoping to seize QZR, his leadership, and the radio station. The plan tasked Cougar Company with clearing Sarowbay. To manage the operation, planners placed a command and control node called the Black Tactical Command Post (TAC) and platoon-sized blocking positions on the ridgeline separating the two objectives. The size of the operation and the physical remoteness of the AO required the marshalling of extensive aviation assets for lift and support drawn from other areas of Afghanistan. An Afghan commando unit advised by an American special operations detachment served as the only official reserve force for STRONG EAGLE III.[2]

To complete his portion of the operation, Captain Tye Reedy, Cougar Company's commander, had only two organic rifle platoons at his disposal to maneuver on Objective LEXINGTON. *No Slack*'s Headquarters and Headquarters Company (HHC) had received its own battle space for the duration of the deployment, a decision that resulted in the transfer of Cougar Company's 1st Platoon to HHC. That move reduced the number of "trigger-pullers" in Cougar Company to approximately 43 men, not including the company HQ element of Captain Reedy, a Radio Telephone Operator (RTO), the Fire Support Officer (FSO), and a Joint Terminal Air Controller (JTAC). The 2d Platoon counted 22 infantrymen and 3d Platoon had 21. Each platoon had two Mk48 machine guns, three M249 Squad Automatic Weapons (SAWs), and three M320 grenade launchers. To provide immediate indirect fire support, Captain Reedy also had a 60mm mortar crew led by Staff Sergeant Jason McDaniel. Although down one platoon, Cougar Company received a number of assets to ensure mission success. First and foremost, the company partnered with a full platoon of 25 Soldiers from the Afghan National Army (ANA). Sensitivity to the local population and the lack of a Government of the Islamic Republic of Afghanistan (GIRoA) presence required that the ANA platoon take the lead while searching for weapons and insurgents in Sarowbay. In addition to the ANA Soldiers, Captain Reedy also received 15 additional personnel that included three interpreters, a military working dog, a combat camera crew, a Low Level Voice Intercept (LLVI) team, and an eight-person squad of Military Policemen. The large number of support personnel led Captain Reedy to remark that, "There were more attachments than I had trigger pullers, it seemed like."[3]

The village of Sarowbay is located in the narrow Ganjgal Valley running from west to east with high ridges on every side (see Figure 1). A small creek bed runs through the valley to the south and west of the village itself. Sarowbay sits in a U-shape with significant elevation changes and terracing within the village itself. Reedy's plan called for Cougar Company to be inserted at a helicopter landing zone (HLZ) called Bee Eater approximately 500 meters from Sarowbay. Both infantry platoons would then traverse the steep mountainsides and move into the valley. The 3d Platoon, led by First Lieutenant Jacob Sass, would then ascend into an overwatch position on the ridgeline to the south of Objective LEXINGTON. Reedy believed that securing the ridgeline would accomplish the dual purpose of keeping it out of enemy hands while affording 3d Platoon a position from which it could cover the entire village with supporting fire. The plan left the ridgeline to the east unoccupied because its distance from most of the *qalats* (clay houses) in Sarowbay meant that it was far less important than other terrain.

Reedy ordered 2d Platoon, led by First Lieutenant Jason Pomeroy, to clear the village. For 2d Platoon's maneuver, Staff Sergeant Jonathan Wray's Weapons squad would bound ahead of the rest of the platoon to assume a second support-by-fire position on the northern ridgeline with its two Mk48 machine guns and other weapons. From its position, Wray's 10-man squad could also monitor the movements of the villagers during the operation. Intelligence estimated the enemy strength in Objective LEXINGTON at two squads of approximately six insurgents each and potentially two additional enemy weapons teams of approximately 10 insurgents as reinforcements. This gave Reedy's company a significant numerical advantage over projected insurgent manpower. Captain Reedy decided to place McDaniel's mortar crew at HLZ Bee Eater, where First Lieutenant Andrew Rinehart's 1st Platoon, Bravo Company, assumed the task of securing both the landing zone and Cougar Company's mortar crew.[4]

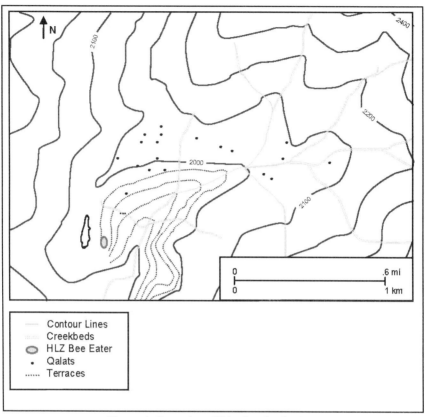

Figure 1. Sarowbay Objective Lexington.

The Operation

Operation STRONG EAGLE III officially began just after midnight on 29 March as CH-47 Chinooks with helicopter gunship escorts began transporting coalition troops to the various HLZs on Objective VIRGINIA. Cougar Company's 3d Platoon with Reedy's headquarters element arrived first on HLZ Bee Eater, approximately half a kilometer to the southwest of Sarowbay. These first landings were delayed by several minutes because gunships had spotted approximately 15 individuals in the vicinity of the landing zone. Later flights by the Chinooks brought in 2d Platoon and the additional assets needed to the clear the village. After the insertion of 1st Platoon, Bravo Company, on HLZ Bee Eater, 3d Platoon set off down the mountainside. Since Sass' platoon had the farthest distance to travel, it left the HLZ first so it could reach its elevated support-by-fire position before dawn. Sass and Specialist Brett Kadlec, the point man, attempted to find a suitable path down the ridge but the terrain proved a significant challenge for the platoon. The steep and rocky landscape made it difficult for Soldiers to get their footing. Often, the only available handholds for Soldiers to grab onto consisted of limbs and roots from the moderately sized trees in the area. The extremely low levels of nighttime illumination in the valley further complicated their movements as the Soldiers' night-vision goggles only partially compensated for the darkness. In addition, each Soldier carried more than 50 pounds of equipment including extra ammunition for their personal weapons as well as ammunition for the platoon's machine guns. The combination of these problems forced Soldiers to noisily "slide on [their] butts" as they moved down the mountainside. Although the straight line distance was perhaps 300 to 400 meters, it took 3d Platoon approximately two hours to complete the move.[5]

Unbeknownst to the Soldiers of Cougar Company, enemy forces quickly detected the American presence and began converging on their positions. The 3d Platoon's Sergeant Dana O'Connor and Specialist Rick Jacobs reported seeing flashlights moving through the trees lower down the ridgeline even before his unit had left HLZ Bee Eater. Meanwhile, officers in orbiting helicopters as part of Vowell's airborne command post first spotted three and then five suspected insurgents leaving from the northwest side of Objective LEXINGTON. Believing one of the unknown men to be Tamidullah, Vowell radioed Reedy and ordered him to take 2d Platoon and intercept the men before they fled the objective area entirely. By this time, Captain Reedy and 2d Platoon had already moved off HLZ Bee Eater heading north toward the southern end of Objective LEXINGTON. Reedy quickly returned to HLZ Bee Eater to drop off the MPs and the other support assets so that the platoon had a better chance to catch up to the fleeing figures. As 2d Platoon set back out in the darkness, AH-64 Apache gunships flying in support of 3d Platoon's movement to the east came off their stations and began to aid Reedy in the interception.[6]

At approximately 0230 hours, 3d Platoon reached the creek bed at the valley floor. Lieutenant Sass did not immediately see a suitable crossing point so he called a short halt to confer with Sergeant Bryan Burgess, the leader of the 3rd squad. After a brief discussion, Sass decided to move northward parallel to the creek bed and search for a better crossing where they could begin their climb up the opposite ridgeline. The Soldiers moved through a series of terraces west of the creek bed and flanked to the left by a densely wooded area. During the movement, the point man, Kadlec, began hearing hushed voices speaking in a foreign language. Kadlec considered the possibility that a portion of their ANA detachment was close by, so he radioed Sergeant Burgess to move to his position. At this point, Kadlec stood at the head of the formation followed by Specialist Dustin Feldhaus and Sergeant Dana O'Connor, the latter of whom was situated five meters behind Kadlec.

Just before Burgess reached Kadlec, gunfire erupted from the treeline. Enemy fighters had climbed into the trees and opened fire almost immediately hitting Burgess, Feldhaus, and O'Connor. The insurgent rounds struck Burgess in his thigh and Feldhaus received multiple gunshot wounds on the left side of his body. The fighters engaged at such short ranges that O'Connor, who took a gunshot wound to his stomach, later found rounds poking through his individual body armor. Several other Soldiers in the ambush zone later reported finding bullet holes through their pant legs that miraculously missed hitting flesh.[7]

Almost instantaneously, someone yelled, "They're in the trees!" and the remainder of the platoon opened fire on the suspected insurgent locations. Even with this knowledge, the poor illumination levels made it difficult for Private First Class Bryan Smith to spot the enemy with his night optics device (NOD). In the darkness, Kadlec helped O'Connor move to an impromptu command post several meters from the ambush site. Meanwhile, Smith focused his efforts on tending to Burgess' wound.[8]

Using its small arms, 3d Platoon quickly neutralized the insurgents who had unleashed the ambush but survivors from this same element of fighters regrouped and were spotted preparing to attack the platoon from the east. In fact, these insurgents appeared in the same area where 3d Platoon intended to establish its support-by-fire position later that morning. Lieutenant Sass, farther to the rear, radioed Captain Reedy that enemy fighters had opened fire on 3d Platoon from positions to the north and east and requested support from 2d Platoon. Simultaneously, Sergeant Jonathan Prins, the Fire Support Non-Commissioned Officer, requested support from Close Combat Aviation (CCA) to secure the area. Immediately, Captain Reedy ended the interception mission and took 2d Platoon back to HLZ Bee Eater for a second time and ordered his men to drop their rucksacks so that they could move more rapidly. In anticipation of their arrival, the leader of 3d Platoon's Weapons Squad, Sergeant James T. Schmidt, ordered his and another squad to take up positions to the west and the southeast of 3d Platoon's position while awaiting the linkup. Sergeant Eric North, the leader of 2d Platoon's 3d Squad, found it very difficult to locate Sass' trail down the ridgeline but he and the rest of the platoon rapidly closed the distance with 3d Platoon by taking a direct route and foregoing standard noise and light discipline. As 2d Platoon moved,

Apache gunships arrived on station to the east of 3d Platoon's position and made several runs firing their 30mm chain guns that blunted the enemy advance and killed 15 insurgents. The 2d Platoon arrived at the ambush site as the gunships attacked, and Sergeant North reported that he could clearly hear the ejected brass cartridges from the helicopters' cannons falling all over the area.[9]

Approximately 10 minutes after the initial ambush while the Apaches were making their gun runs, Specialist Rick "Doc" Jacobs arrived to tend to the three wounded Soldiers. Burgess drifted in and out of consciousness but Jacobs managed to stop the bleeding from his thigh wound. Feldhaus, despite having sustained several gunshot wounds, remained conscious and alert throughout the brief engagement. O'Connor's stomach wound appeared stable and non-critical. Unfortunately, it took considerable time for the MEDEVAC helicopters to arrive at the scene. Before the operation, Cougar Company was briefed that MEDEVACs could reach their objective in approximately five minutes from Forward Operating Base (FOB) Joyce. Instead, it took nearly 30 minutes before the MEDEVAC helicopter arrived to evacuate the wounded. As the now linked up 2d and 3d Platoons provided security, the UH-60 MEDEVAC hoisted the three casualties aboard and immediately set off for the hospital.[10]

With the tree line cleared and the casualties evacuated, Captain Reedy reconstituted the company at approximately 0400 hours and briefly conferred with Lieutenant Sass to decide how to proceed with the mission. Sass' 3d Platoon had only 18 riflemen remaining, and the casualties significantly reduced its capability to hold its planned overwatch position on the southern ridgeline. Reedy also believed that the loss of three men, including Sergeant Burgess, whose experience as a squad leader made him a key leader in the unit, would have negative psychological and morale effects upon the remaining Soldiers. With those two factors in mind, Reedy altered his initial plan for clearing Sarowbay. He still planned for 2d Platoon to clear the village with Staff Sergeant Wray's weapons squad providing local support-by-fire on the ridgeline above Sarowbay. The biggest change to the plan affected 3d Platoon. The unit would now provide local-support-by-fire from the ridgeline to the southwest of Sarowbay during the initial phase of clearances. Reedy believed that the risk of leaving an under-strength unit in isolation outweighed the threat posed by potential enemy control of the surrounding terrain.[11]

With the new plan in place, Captain Reedy and his men set out once again for Objective LEXINGTON. Cougar Company marched back up the ridgeline and recovered both their dropped rucksacks and support units.

The military working dog and its handler stayed on the LZ due to concerns over the security situation following the ambush of 3d Platoon and the ability of the dog crew to keep up with the fast moving company as it maneuvered towards the objective. By 0700 hours, the company reached a position above Sarowbay. The 3d Platoon assumed a support-by-fire position near a graveyard below HLZ Bee Eater while 2d Platoon and the support assets set about clearing the western end of Sarowbay. When dawn broke on the morning of 29 March, Cougar Company was poised to clear its primary objective.[12]

As the initial clearing operations proceeded during the morning, the women and children of Sarowbay set about accomplishing their daily chores seemingly oblivious to the fighting that occurred just a few hours prior. The Soldiers of 2d Platoon found neither significant caches of weapons nor military aged males, who had presumably fled long before the arrival of American Soldiers. They did, however, find much evidence confirming an insurgent presence in Sarowbay. Most significantly, the Soldiers discovered that the local civilians tuned their radios to the frequency of Tamidullah's propaganda radio station that the scout platoon of TF *No Slack* had found and destroyed on the northern ridgeline that morning. One house in particular yielded significant items of interest. Intelligence had located the home that they believed belonged to Tamidullah or at the very least, functioned as his hiding place. When searching the *qalat*, Soldiers discovered an elderly Afghan later identified as Tamidullah's father. They also located some weaponry in the structure, including Light Anti-Armor Weapons (LAWs) and other US issued equipment carried by the Marines killed in the Ganjgal Valley in 2009. To the insurgents, these reminders of battlefield success conferred respect in war torn Afghanistan where victory against foreigners constituted a means of attaining prestige and social mobility.[13]

As the morning progressed, 2d Platoon cleared the western end of the village and began circling the U-shaped turn towards the *qalats* on the south side of the northern ridgeline. The village of Sarowbay contained between 150 and 200 *qalats,* enough to potentially house hundreds of Afghans, although it was not clear how many were occupied at any given time. Additionally, Reedy and his men had discovered earlier in their deployment that clearing operations in Afghanistan took much longer than similar operations in Iraq. Reedy therefore decided that morning to begin scouting potential locations for his troops to rest during the night. One elderly Afghan woman offered to let Lieutenant Pomeroy and 2d Platoon stay in her home for the evening but Reedy decided to press on to a series of *qalats* further to the east.[14]

Once 2d Platoon reached this group of *qalats*, the Soldiers of 3d Platoon left their support-by-fire position near the graveyard to the southwest and began moving toward *qalats* at the bottom of the valley beneath those previously searched by 2d Platoon earlier in the morning. While these searches were underway, Reedy received a radio message from Major William Rockefeller, the TF *No Slack* executive officer, asking him to turn on his Thuraya satellite phone so they could have a conversation off the battalion communication channels. Dreading what Rockefeller would tell him, Reedy picked up the phone and was informed that Sergeant Burgess died from the wounds he sustained in the morning ambush. Reedy decided that the men of 3d Platoon deserved to know this news as quickly as possible, so the company commander, his JTAC, and his RTO, left 2d Platoon's position and went to the valley floor at approximately 1000 hours. After gathering 3d Platoon together, Reedy told them of the death of their respected squad leader. "You could immediately feel the impact of losing him," Reedy said adding, "I told the men they had five minutes to pray, think, cry, or punch a wall."[15]

At this point in the morning, just before 1100 hours, the men of Cougar Company were separated into three groups. The 3d Platoon was at the bottom of the valley, 2d Platoon occupied a multi-room *qalat* above and to the east of 3d Platoon's position, and Wray's weapons squad was farther up the northern ridgeline (see Figure 2). The 2d Platoon's strongpoint *qalat* was larger than nearly all of the other structures in the village but the structure was built directly into the side of the ridgeline and had no clear view northward up the ridge. Wray's squad had shifted positions throughout the morning to maintain their support-by-fire position over 2d Platoon and now they focused their attention on protecting the flank of the strongpoint *qalat*. To accomplish this task, Wray's squad took up a position on the ridgeline above and to the east to guard the trail through the objective.[16]

The clouds gathering in the skies above posed yet another problem for Captain Reedy. The mountains of eastern Afghanistan caused erratic weather patterns with significant temperature swings and the rapid appearance of thick clouds and precipitation, particularly during the afternoons. The mountainous terrain interfered with communications systems and the bad weather threatened to force Close Air Support (CAS) and CCA out of the area thus depriving Cougar Company of one of its most important sources of fire and intelligence, surveillance, and reconnaissance (ISR) support. Reedy, still with the men of 3d Platoon at the bottom of the valley, decided to consolidate the company at 2d Platoon's position to prevent the enemy from cutting off a portion of his force if weather conditions deteriorated further. Unfortunately for 3rd Platoon, the terrain prevented the men from moving directly towards 2d Platoon's *qalat*. Instead, the terraces within the village forced 3d Platoon to move in a U-shaped pattern thus stretching the walking distance between the two halves of the company to approximately 500 meters.[17]

At approximately 1100 hours, Cougar Company's CAS and CCA assets left their stations over Sarowbay because of the drop in visibility. The thickening clouds brought heavy rain and hail into the valley. This same weather system affected every TF *No Slack* position on Objective VIRGINIA from Alpha Company's position eight kilometers to the north to HLZ Bee Eater at the southern end of the objective. With the departure of the aircraft, Cougar Company's fire support was reduced to its 60mm mortar on HLZ Bee Eater and the 105mm howitzers stationed at FOB Joyce approximately nine kilometers to the west. Insurgents, many of whom likely moved out of Sarowbay once they detected the American presence, had covertly occupied fighting positions on the surrounding ridgelines over the previous several hours. The enemy had monitored the movements of Cougar Company from the ridgelines and clearly understood that the departure of the aviation assets left the Americans exposed. Taking advantage of this sudden drop in the Americans' fire support, they decided to strike.[18]

Almost immediately after the aircraft left the area, Reedy and the men of 3d Platoon heard sporadic fire to the south aimed at HLZ Bee Eater as they were moving toward 2d Platoon's position. Knowing that the enemy would soon direct their fire at the exposed 3d Platoon as it moved through open ground, Reedy and his men broke into a trot or the "airborne shuffle" as they described it, in hopes of reaching the strongpoint before that could happen. However, by the time 3d Platoon traversed roughly half of the 500-meter distance, enemy insurgents had opened fire with RPK and PKM machine guns on the platoon from multiple locations on the ridgelines to the north, east, and south. The volume of fire suggested the presence of far greater numbers of insurgents than previously anticipated. Prior to the start of the operation, intelligence estimates pegged enemy strength in and around Objective LEXINGTON as approximately squad-sized. Unfortunately for Cougar Company, the actual number of insurgents on the objective likely numbered between 100 and 150 fighters. Given Cougar Company's losses sustained during the ambush before dawn, the insurgents now held an approximate two to one advantage over the Americans in numerical fighting strength.[19]

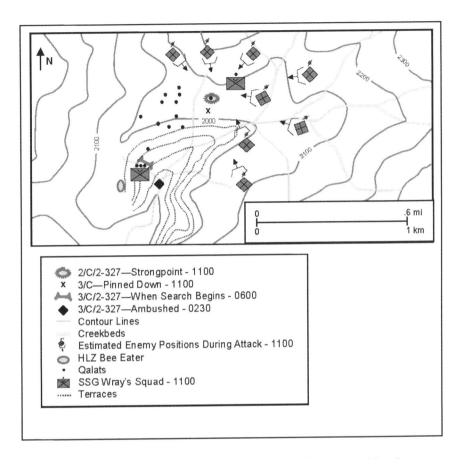

Figure 2. Objective Lexington, 29 March 2011, Terrain features and timelines.

When the enemy opened fire, 2d Platoon's Corporal John R. Nesse and his team left the cover of the *qalat* and laid down suppressing fire. With Nesse's team providing cover, Reedy and the trailing third of 3d Platoon picked up the pace and managed to scurry up the ridgeline into the strongpoint *qalat*. Meanwhile, the leading two-thirds of 3d Platoon had pushed forward on the trail forcing them to take cover behind a two-meter high rock wall 100 meters due south and two terraces below the strongpoint. Sergeant Joshua Bostic, along with Private First Class Jeremy Faulkner and Specialist Joseph Kintz, positioned at the eastern end of the wall and returned fire against the enemy positions to the east. Meanwhile, toward the rear of the group of Soldiers strung out along the rock wall, Lieutenant Sass found himself almost completely exposed to enemy fire. He, Specialist Jacobs, and Specialist Matheson moved back a few meters to a small defilade position in hopes of finding more cover. The enemy fire, extremely heavy at this point, poured onto this position and left the bulk of 3d Platoon with few options.[20]

At the strongpoint *qalat*, the Soldiers of 2d Platoon worked to return fire against the many enemy positions. With all of 2d Platoon, the ANA Soldiers, and the support assets inside the structure, crowding quickly became an issue. Making matters worse, the *qalat* only had one window from which the Soldiers could effectively return fire. From this window facing the south, 2d Platoon's Soldiers returned fire as best they could but many of the known enemy positions were outside of the field of fire offered by the window. Because the trapped elements of 3d Platoon lay to the south, the Soldiers in the *qalat* directed their fire carefully in an attempt to avoid any friendly-fire incidents. Unfortunately, these precautions further limited the platoon's freedom of action. In spite of this, the Soldiers in the *qalat* managed to restore some measure of fire superiority over the insurgents, thanks due in part to Nesse and his team leaving the cover of the *qalat* to return fire. Inside the *qalat*, Specialist Michael Patterson fired his M4 rifle and M203 grenade launcher through the window at enemy positions on the ridgeline to the southeast. The volume of fire increased when 2d Platoon brought an Mk48 machine gun into action.[21]

For the trapped Soldiers of 3d Platoon, the situation had worsened. The enemy PKM positions to the north, south, and east kept the Soldiers under heavy fire and within a few minutes the enemy increased the pressure when another machine gun suddenly opened fire from the west. Lieutenant Sass assessed his position as no longer tenable and began plotting his movement to the strongpoint. By this time, 2d Platoon had regained fire superiority from the *qalat* above. Over the radio, Sass told his men to wait for the cloud cover to thicken further to provide cover for a rapid movement up the ridgeline. When the weather cooperated a short time later, a group of Soldiers including Sass, Kadlec, Jacobs, Matheson, and Sergeant Clint Lyons, made a mad dash to the strongpoint. All of the men from this group reached the *qalat* safely but their relief quickly dissipated when they realized that several Soldiers including Bostic, Kintz, Faulkner, Smith, and Specialist Enrique Gonzalez remained trapped behind the rock wall farther down the slope. For some reason, this element had not received the order to move to the building.[22]

With most of the platoon now in the strongpoint, the remaining Soldiers trapped outside took the brunt of enemy fire. As the men of 3d Platoon sat crouched behind the wall, a bullet which his fellow Soldiers initially believed hit his arm, struck Faulkner and expended its energy against his armor plating. Moreover, Bostic sent a radio message to Reedy reporting that Faulkner's wound was not serious. Unfortunately, the enemy round actually struck just above Faulkner's plating and pierced his chest. Kintz and Bostic realized that Faulkner had suffered a serious wound and tried to provide medical care while still exposed to enemy fire. Despite their best efforts, Faulkner died on the battlefield barely 30 seconds after being hit. While Bostic tried to perform first aid on Faulkner, an enemy round struck him in the buttocks.[23]

The situation in the strongpoint had improved somewhat while 3d Platoon endured its hellish ordeal on the terraces below. The Soldiers in the *qalat* continued to bring their own weapons to bear against insurgent positions. Enemy rounds hit and sometimes entered the *qalat*, but no members of 2d Platoon suffered any wounds. These rounds, however, made it easier to determine the enemy's positions by allowing First Lieutenant Steven Craig, the company's Fire Support Officer, to perform some back azimuth calculations. They relayed this information on the enemy positions via TacSat (tactical satellite) back to Sergeant McDaniel's mortar crew on HLZ Bee Eater which began dropping rounds on the insurgents. The mortar, however, remained the company's sole source of fire support because communications difficulties prevented Reedy from making continuous contact with FOB Joyce. In any case, the beleaguered company had no access to the 105mm howitzers at the FOB because the sheer volume of fire mission requests coming from every unit in TF *No Slack* had overwhelmed the battery.[24]

While all of these events transpired below, Staff Sergeant Wray's squad found itself undergoing its own trial farther up the ridgeline. Wray had 10 Soldiers under his command and his squad carried two Mk48 machine guns, a SAW, and M14 sniper rifle. At his squad position to the northeast of 2d Platoon's strongpoint, Wray had deployed his machine guns to cover the eastern and southern approaches respectively with the SAW protecting the rear of the squad's position to the west. When the insurgents opened fire at approximately 1100 hours, Wray initially ordered his squad to maintain concealment so that they could better determine the origins and the targets of enemy fire. He quickly realized that his squad's position was receiving heavy fire from the tree line across the valley to the east/southeast. The squad then began responding to the enemy fire with Wray moving between his Soldiers' positions to help them more accurately aim their fire. The insurgents proved extremely aggressive, maneuvering in the open to take clear shots at the squad, causing Wray to shift his men's position to maximize their cover. Wray attempted to radio Lieutenant Pomeroy to share information, but he encountered the same communication problems bedeviling the rest of Cougar Company and could only maintain sporadic contact.[25]

During the firefight, the squad's designated marksman, Specialist Frederick Hutterli, took notice of a spotter a few hundred meters away and across the valley gesturing to his rear and pointing directly at the squad's position. Although not a formally trained sniper, Hutterli relished the chance to fire on a target from long range and sighted his M14 on the spotter. Just as he fired, Wray dropped a magazine near Hutterli, causing him to miss the target. Fortunately, the shot still had a deterrent effect and the spotter slipped inside a nearby *qalat* and gave up his mission.[26]

At approximately 1140, the SAW gunner, Specialist Travis Bland, noticed a pair of enemy fighters manning a machine gun and trying to outflank the squad's position from farther up the ridgeline. After receiving word from Lieutenant Pomeroy that there were no friendlies anywhere on the ridge, Wray asked for two volunteers to help him neutralize the machine gun. Specialists William Dempsey and Matthew Neal answered the call and the three Soldiers began maneuvering towards the enemy position with Bland covering them with his SAW. During the maneuver, a second machine gun on a nearby hill opened fire, pinning Dempsey and Neal down. With Bland trying to suppress both machine guns, Wray closed on the first machine gun. He never laid eyes on the enemy position but he lobbed a grenade in the vicinity which, after it exploded, caused the gun to stop firing. The most significant action occurred when Wray's squad brought an enemy PKM crew under fire. Lieutenant Pomeroy informed Wray via radio that at least eight insurgents were on the ridgeline near Wray and ordered him to gather his squad and move toward the 2d Platoon strongpoint. As Wray carried out these movements, the second insurgent machine gun crew continued to pour fire onto the squad.[27]

While Wray protected 2d Platoon's northern flank, the Soldiers inside the strongpoint *qalat* in the village worked to relieve 3d Platoon's Bostic, Gonzalez, Kintz, and Smith farther down the ridgeline. As the Soldiers organized a rescue mission, a large blast knocked everyone to the ground. The *qalat* occupied by the old woman who had tried to persuade Lieutenant Pomeroy to bunk there that night exploded when an improvised explosive device with a cell phone trigger detonated inside the structure. Fortunately, the blast was too distant to cause any Coalition casualties. Everyone picked themselves up and resumed their focus on their besieged comrades.[28]

Lieutenant Pomeroy organized a small rescue team comprised of himself, Specialist Nathan Allen, Specialist Jordan Anderson, and Nesse to provide cover so that the trapped men from 3d Platoon could scramble up the ridgeline to safety. Pomeroy's rescue team darted from the strongpoint and made for a group of smaller *qalats* to the west but his group quickly ran into trouble. Barely 15 meters outside the door, enemy fire struck Allen when a round penetrated a small rock wall that he tried to use as cover, piercing Allen's arm and lodging in his liver. Seeing Allen wounded, Pomeroy halted the rescue mission and focused the team's efforts on saving the wounded Soldier. Pomeroy tore apart the rock wall to reach Allen and Pomeroy and Nesse dragged their wounded comrade to safety. In order to allow Pomeroy's men to return to the cover of the strongpoint, Sergeant James Nichols gathered two MPs and left the safety of their *qalat* to provide covering fire. In spite of his wound, Allen moved quickly under his own power back to the strongpoint. His uninjured comrades also made it back to the strongpoint safely.[29]

Once the team returned to the *qalat*, Reedy made yet another wrenching decision and cancelled any further rescue missions until CAS and CCA could help the company suppress enemy fire surrounding the village. The situation outside was too dangerous and the hail of insurgent gunfire too heavy for Soldiers to risk exposure to an enemy force superior in both numbers and tactical position. In spite of these problems, the men of Cougar Company worked systematically to suppress enemy fire emanating from locations that could also threaten Bostic's group below. Bostic, Gonzalez, Kintz, and Smith had by this time endured a constant fusillade of fire focused on their position for the better part of an hour. With cover provided by the men in the *qalat*, the four Soldiers darted from their meager cover and made their way as a group up the ridgeline to the strongpoint. Unfortunately, the need for haste prevented the group from taking Faulkner's body or any of his gear with them.[30]

With the surviving members of the company now consolidated in the *qalat*, recovering Faulkner's body became their priority. At this point in the fight, Wray's squad had completed their move down the ridgeline and quickly cleared and occupied a *qalat* 100 meters to the west of the 2d Platoon strongpoint. Only now as the communication problems began to ease did Wray learn of the ordeal endured by the rest of Cougar Company. He volunteered his squad to try and recover Faulkner's body but Reedy's edict against further rescue missions stood. The men of Cougar Company continued to defend their position and wait. Finally, after more than an hour of rain and hail, the cloud cover began to dissipate enough for CAS and CCA to return to their stations over Sarowbay. Almost immediately, CCA initiated gun runs against enemy positions scattered across the various ridgelines surrounding Sarowbay.

With an exponential increase in the amount of fire support, the men of Cougar Company formed a new team to retrieve Faulkner's body.

Specialist Kintz left the strongpoint *qalat* and provided suppressing fire for the rest of the team that included Nesse, Sergeant James Nichols, and Patterson. The four made their way down to the terrace where Patterson assumed the task of recovery. In spite of the attacks from the aircraft overhead, the enemy continued to fire on the American Soldiers. Patterson briefly searched the area where 3d Platoon had found cover to retrieve any essential equipment left behind before the different groups of Soldiers made their mad dashes for the strongpoint. After completing this first task, Patterson picked up Faulkner's body and, while protected by the rest of the team's suppressing fire, made his way back to the strongpoint.[31]

Soon after the recovery of Faulkner's body, a MEDEVAC helicopter finally arrived to bring the wounded off the battlefield. Patterson brought Faulkner's body to the helicopter and then helped the wounded Allen climb aboard. With Faulkner and Allen extracted, the Soldiers of Cougar Company could remain sheltered in their *qalats* until the security situation improved outside. Enemy fire continued throughout the day but slackened enough for Sergeant Wray and another Soldier to collect ammunition from the strongpoint *qalat*. Noticing the crowded conditions in the strongpoint, Wray decided to keep his squad at their current location and Reedy sent a fire team led by Sergeant Nichols to reinforce their position. The enemy still continued to press into the evening hours and one enemy element ventured close enough to the *qalat* that the JTAC called in a bomb strike that neutralized the insurgent force.[32]

That night, Captain Reedy spoke with Lieutenant Colonel Vowell about the status of Cougar Company's mission. The clearing operations that morning had only accounted for a roughly 50 of the *qalats* in Sarowbay and Reedy estimated that between 50 and 100 dwellings remained. The status of 3d Platoon's men also weighed heavily in Reedy's thoughts. Third platoon had lost three killed and two wounded and sustained all of the company's casualties on Objective LEXINGTON. In addition to these physical losses, at least two of the other Soldiers showed signs of significant mental strain. The combined effect of the predawn ambush and the midday ordeal had rendered the platoon unable to function as a maneuver element for the remainder of the operation. The firefights had also left the company short of water and ammunition. Reedy had two options. These were to continue the mission with his company crucially short on combat power or ask for reinforcements that could complete the clearance of Sarowbay. With all of these factors in mind, Reedy made a decision that he characterized as the "hardest but easiest" choice. He requested that Vowell dispatch additional forces to Objective LEXINGTON. Early on the morning of 30 March, the battalion commander sent Charlie Company, 1-327 IN also nicknamed Cold Steel, to HLZ Bee Eater. Once the helicopters completed their lifts, Cold Steel moved down and secured the western end of Sarowbay more than doubling the number of Coalition forces and promising to improve the security situation in the village.[33]

Even though the new units dramatically reduced Cougar Company's burden, the second day still proved dangerous. At 0700 hours, Reedy learned that Feldhaus had succumbed to the wounds sustained during the ambush coming off of HLZ Bee Eater. Just as he had done when notified of Burgess' death, Reedy gathered the men in the strongpoint *qalat* and gave them several minutes to reflect. That morning, Cougar Company fortified their strongpoint by building makeshift walls with rocks and sandbags. Soldiers in the *qalat* heard a mysterious noise behind a closed door and discovered several women and children hiding in the small space behind it. These villagers had stayed in the house throughout the chaotic events of the previous day and managed to escape detection. After the lone female military policeman searched the women and children, they were allowed to go about their daily routines unobstructed.[34]

Like the first day on Objective LEXINGTON, the weather began to turn against Cougar Company that afternoon. Heavy clouds reduced visibility to less than 10 meters and brought rain and hail that drove aviation assets away from the area. This provided a new window for enemy fighters to attack. Unlike the first day however, no American unit lay completely exposed to enemy fire. Since Wray's flanking element no longer guarded the northern flank of the strongpoint *qalat*, the Soldiers of Cougar Company fired and lobbed grenades up the ridgeline to disrupt enemy movements. They hoped that using their stock of grenades would conserve ammunition for Sergeant McDaniel's mortar on HLZ Bee Eater, their most consistent fire support element throughout the operation. In fact, the rules of engagement for the mortar expanded on the second day when the men of Cougar Company noticed enemy fighters occupying *qalats* at the far eastern end of the village. After some of McDaniel's mortar rounds struck near the occupied *qalats*, the insurgents responded by trying to gather civilians in *qalats* occupied by their spotters. By doing so, insurgents hoped that Americans would withhold their fire rather than risk inflicting civilian casualties. In any case, the mortar rounds caused these insurgents to stop firing on Coalition forces for the remainder of the engagement.[35]

Later that afternoon, weather patterns held course and the cloud cover dissipated, allowing for the return of aviation assets. That night, Cougar

Company received even more reinforcements when an ANA commando element arrived on the objective to resume clearance operations to the east of the strongpoint. On the morning of the third day, Reedy linked up with the leaders of the commandos to give them a situation report and information on suspected enemy positions on the surrounding ridges. As the Afghan Commandos moved to the east, 2d Platoon left the strongpoint *qalat* and assumed a support-by-fire position during the clearances. At some points, the presence of CAS and CCA allowed Captain Reedy to dispatch small patrols into the surrounding area. The situation did not allow for an extended battle damage assessment but the men of Cougar Company discovered several destroyed enemy positions that demonstrated the accuracy of their CAS, CCA, and indirect fires. Although the exact number of enemy casualties remained unknown since the insurgents often removed their dead to hide their numbers, Cougar Company estimated that they inflicted between 50 and 100 casualties on the enemy surrounding Sarowbay. Cougar Company also detained two military aged males and sent them back to FOB Joyce for processing.[36]

These missions allowed for some breaks in Cougar Company's routine but the bulk of the unit spent most of their time from 30 March into the evening of 1 April occupying the strongpoint. The presence of the support assets kept conditions in the *qalat* crowded and the ANA unit present for the mission provided little additional combat power. The ANA proved extremely useful during the initial clearing operations but when the weather turned poor and the insurgents attacked, they had done nothing to defend the coalition positions in Sarowbay. In fact, during the afternoon firefights, the Soldiers of Cougar Company almost invariably appropriated ammunition from their ANA counterparts because, as Corporal Nesse put it, "we figured we could use their ammo better than they could." Cougar Company received only intermittent resupply during the four days on the objective, and the US Soldiers began to view the Afghans as a drain on their stocks of water and ammunition.[37]

Finally, on the night of 1 April, after four days out in the field, TF *No Slack*'s mission changed. What had started as an effort to neutralize QZR's command and control structure in Kunar Province turned into an active manhunt for prominent insurgent leaders. As a result, the focus of the action shifted southward and Lieutenant Colonel Vowell retasked most of the company to accommodate the changing mission requirements. The Soldiers of Cougar Company had recovered from the ordeal of the first day and became stir crazy within the strongpoint *qalat*. As Reedy himself put it, he was willing to do anything to "get me the f--- out of this house."[38]

Despite a significant loss of effective strength during the previous four days of fighting, Cougar Company soldiered on through the remainder of STRONG EAGLE III. After the company walked back to HLZ Bee Eater, CH-47s transported the Soldiers to HLZ Mallard which overlooked the village of Shirugay, codenamed Objective NORFOLK-EAST. For the remaining four days of the operation, Cougar Company provided overwatch while the ANA commando detachment cleared the village. At no point did the men of Cougar Company make contact with the enemy. On the night of 5 April, Operation STRONG EAGLE III came to an end and the Soldiers from the company boarded the Chinooks and returned to FOB Joyce. By the time they returned, Sergeant O'Connor had resumed his duties after his stay in the hospital. Over the next three weeks, the Soldiers of Cougar Company ended their deployments and left Afghanistan.[39]

Aftermath and Conclusions

Reedy and his men believed that STRONG EAGLE III destroyed QZR's command and control node and made border crossing by enemy fighters much more dangerous. Unfortunately for Cougar Company, their imminent departure from Afghanistan meant that they would not see the results of their work. Reedy hoped that the lethal and aggressive nature of the operation would serve as a model for "continuous disruption" and shape future operations conducted in the border regions of Afghanistan. To Reedy, only the destruction of insurgent networks would allow the government of Afghanistan to expand its influence and ensure the security of the local population.[40]

The fight for Objective LEXINGTON clearly highlighted the difficulties of operating in mountainous eastern Afghanistan. More than any other single factor, terrain influenced the conduct of the operation. The rugged natural landscape significantly slowed troop movement and the high altitudes caused the sudden onset of weather that eliminated the American advantage in aerial fire support for hours at a time. The location of villages, such as Sarowbay, at the bottom of valleys dominated by high ridgelines presented military leaders with operational and tactical challenges. Captain Reedy tried to forestall enemy control of the high ground by establishing one of his platoons in an elevated support-by-fire position. The enemy, however, inadvertently disrupted this plan by successfully executing the pre-dawn ambush. Inside the village, the man-made terraces complicated movements between *qalats*. The loss of Private Faulkner during the midday firefight on 29 March demonstrated the extreme to which the mountains and weather can effect military operations but Cougar Company successfully employed their own small arms to neutralize the enemy's positional advantage on the ridgelines long enough for the remainder of 3d Platoon to reach cover.

Cougar Company showed remarkable resiliency and adaptability throughout STRONG EAGLE III in the face of a tough and determined enemy which knew the terrain well. Much of this success came from the able leadership of the company at all levels. At several points on the first day of the operation alone, circumstances dictated complete revisions to Cougar Company's plan for clearing Objective LEXINGTON. When the early morning ambush of 3d Platoon reduced its combat capability, Reedy quickly changed his plan for clearing the village that took this into account yet still provided 2d Platoon with the appropriate support-by-fire. The weather completely disrupted the clearance of Sarowbay and resulted in more harrowing moments for much of 3d Platoon. The unknown strength of the enemy in the Ganjgal area disrupted the fires priority for the entire battalion and resulted in Cougar Company not receiving the desired artillery support at crucial moments of the engagement. As Lieutenant Craig joked, "We're still waiting on word from some suppression missions." In spite of these complications, Cougar Company successfully established a strongpoint in the village, defended it from determined enemy attacks, and provided the necessary suppressing fire to allow the exposed Soldiers of 3d Platoon to reach safety. Reedy himself displayed remarkable loyalty and respect twice by personally informing his men when one of their comrades had died. He further displayed courage and integrity when he requested reinforcements after recognizing that the mission had grown too large for his under strength and battle weary company to complete alone. Through it all, even after four days of contact with the enemy, Cougar Company accepted the change in mission and remained in the field for the duration of the operation.[41]

The Soldiers of Cougar Company believed that Operation STRONG EAGLE III proved the hardest of all of the operations they conducted during a very active deployment. The company made a significant contribution to the operation's success but realized that they had survived a dangerous ordeal. As Corporal Nesse said, "I saw more action in the [Ganjgal] valley over ten days than I did in an entire year. I've seen, you know, things that will put you in the fetal position and make you cry. I saw some amazing things in ten days."[42]

Notes

1. Lieutenant Colonel Joel B. Vowell and Command Sergeant Major Chris Fields, interview by Ryan Wadle, Combat Studies Institute, Fort Leavenworth, KS, 11 July 2011, 2-5, 50.

2. Major Eric Anderson, Major Phillip Kiniery, Major William Rockefeller, interview by Ryan Wadle, Combat Studies Institute, Fort Leavenworth, KS, 11 July 2011, 4-15.

3. Captain Tye Reedy, quoted in Charlie Company, 2-327 IN, group interview by Ryan Wadle, Combat Studies Institute, Fort Leavenworth, KS, 14 July 2011, 6. Captain Tye Reedy, "Operation Strong Eagle III – The "Other Objective": The Fight on Objective Lexington, Cougar Company, 2d BN, 327th IN Regiment," unpublished paper, 2011, 3; Captain Tye Reedy, e-mail to Ryan Wadle, Combat Studies Institute, 12 September 2011.

4. Reedy, "The Other Objective," 2-3; Staff Sergeant Jonathan Wray, interview by Ryan Wadle, Combat Studies Institute, 23 September 2011, 11-3.

5. Specialist Brett Jacobs, quoted in 3d Platoon, Charlie Company, 2-327 IN, group interview by Ryan Wadle, Combat Studies Institute, Fort Leavenworth, KS, 14 July 2011, 6. 3/C/2-327, group interview, 5-7; Wray, interview, 7-8; C/2-327, group interview, 20-1; Reedy, e-mail, 12 September 2011.

6. 3/C/2-327, group interview, 17; Reedy, group interview, 22; First Lieutenant Jacob Sass, "Tactical Decision Scenario," unpublished paper, 2011, 1.

7. C/2-327, group interview, 23, 26; 3/C/2-327, group interview, interview, 9-10.

8. Sergeant Dana O'Connor, quoted in 3/C/2-327, group interview, 9. 3/C/2-327, group interview, 11, 13.

9. 3/C/2-327, group interview, 10-14; Reedy, "The Other Objective," 4; C/2-327, group interview, 24-31; Wray, interview, 9.

10. 3/C/2-327, group interview, 12, 14; Reedy, "The Other Objective," 4; C/2-327, group interview, 29, 31.

11. C/2-327, group interview, 32; Reedy, "The Other Objective," 4; Captain Tye Reedy, e-mail to Ryan Wadle, Combat Studies Institute, 26 August 2011.

12. C/2-327, group interview, 32; 3/C/2-327, group interview, 16.

13. Wray, interview, 13-4; C/2-327, group interview, 32-3; Joseph Holliday, interview by Ryan Wadle, Combat Studies Institute, Fort Campbell, KY, 13 July 2011, 12-5.

14. C/2-327, group interview, 18-9, 34-5, 60.

15. Reedy, "The Other Objective," 4. C/2-327, group interview, 36-8; 3/C/2-327, group interview, 16.

16. C/2-327, group interview, 36, 44.

17. C/2-327, group interview, 38-9.

18. C/2-327, group interview, 34, 38.

19. C/2-327, group interview, 39-40, 61.

20. Narrative to Accompany the Award of Bronze Star Medal with Valor to Specialist John R. Nesse, no date; Reedy, e-mail, 12 September 2011; C/2-327, group interview, 40-1; 3/C/2-327, group interview, 21; Reedy, "The Other Objective," 5.

21. C/2-327, group interview, 43; Narrative to Accompany the Award of Bronze Star Medal with Valor to Specialist Michael G. Patterson, no date; Captain Tye Reedy, e-mail message to Ryan Wadle, Combat Studies Institute, 31 August 2011.

22. C/2-327, group interview, 41; 3/C/2-327, group interview, 22.

23. C/2-327, group interview, 42; Narrative to Accompany the Award of Bronze Star Medal with Valor to Specialist Joseph M. Kintz, no date; 3/C/2-327, group interview, 23; Reedy, e-mail, 26 August 2011.

24. C/2-327, group interview, 52; Reedy, "The Other Objective," 5.

25. Wray, interview, 2-3, 16-7, 24.

26. Wray, interview, 23-4.

27. Narrative to Accompany the Award of Army Commendation Medal with Valor to Private First Class Travis C. Bland, no date; Wray, interview, 18-21, 25.

28. C/2-327, group interview, 49.

29. C/2-327, group interview, 46, 48; Narrative to Accompany the Award of Bronze Star Medal with Valor to Specialist Nathan M. Allen, no date; Narrative to Accompany the Award of Bronze Star Medal with Valor to Nesse; Reedy, "The Other Objective," 5; Narrative to Accompany the Award of Bronze Star Medal with Valor to Sergeant James W. Nichols.

30. C/2-327, group interview, 47-8; Reedy, e-mail, 26 August 2011.

31. Wray, interview, 22, 26-7;C/2-327, group interview, 48; 3/C/2-327, group interview, 25; Narrative to Accompany the Award of Bronze Star Medal with Valor to Kintz; Narrative to Accompany the Award of Bronze Star Medal with Valor to Nesse; Narrative to Accompany the Award of Bronze Star Medal with Valor to Nichols; Narrative to Accompany the Award of Bronze Star Medal with Valor to Patterson.

32. 3/C/2-327, group interview, 26; Narrative to Accompany the Award of Bronze Star Medal with Valor to Patterson; C/2-327, group interview, 52-3; Wray, interview, 28.

33. Reedy, "The Other Objective," 7. Reedy, e-mail, 12 September 2011; C/2-327, group interview, 54-5; Tye Reedy, e-mail, 26 August 2011.

34. C/2-327, group interview, 56, 60; 3/C/2-327, group interview, 27.

35. Reedy, "The Other Objective," 7; C/2-327, group interview, 62-3, 65-7; Wray, interview, 30-2.

36. C/2-327, group interview, 55-8, 83; Reedy, "The Other Objective," 8.

37. Corporal John Nesse, quoted in C/2-327, group interview, 15. C/2-327, group interview, 12-5.

38. Reedy, quoted in C/2-327, group interview, 68.

39. C/2-327, group interview, 68-9, 73; 3/C/2-327, group interview, 34-5.

40. Reedy, quoted in C/2-327, group interview, 91-4.

41. First Lieutenant Steven Craig, quoted in C/2-327, group interview, 91.

42. Nesse, quoted in C/2-327, group interview, 87.

Glossary

ANA	Afghan National Army
ANP	Afghan National Police
ANSF	Afghan National Security Forces
ARF	Aerial Reaction Force
ASR	Alternate Supply Route
ATPIAL	Advanced Target Pointer Illuminator Aiming Light
AWG	Asymmetric Warfare Group
BCT	Brigade Combat Team
C2	Command and Control
CAT	Civil Affairs Team
CAV	Cavalry
CCA	Close Combat Attack
CERP	Commander's Emergency Response Program
CFSOCC	Combined Forces Special Operations Component Command
CJSOTF	Combined Joint Special Operations Task Force
COIN	Counterinsurgency
COP	Combat Outpost
CP	Command Post
CROWS	Common Remotely Operated Weapon Station
FA	Field Artillery Regiment
FO	Forward Observer
FOB	Forward Operating Base
HLZ	Helicopter Landing Zone
HVT	High Value Targets
IED	Improvised Explosive Device
IJC	International Joint Command
IN	Infantry
IOTV	Improved Outer Tactical Vest
ISAF	International Security Assistance Forces
ISR	Intelligence, Surveillance and Reconnaissance
JDAMS	Joint Direct Attack Munitions
JLENS	Joint Land Attack Cruise Missile Defense Elevated Netted Sensor

JRTC	Joint Readiness Training Center
JTAC	Joint Terminal Attack Controller
KIA	Killed In Action
LP	Listening Post
LZ	Landing Zone
M-ATV	Mine Resistant Ambush Protected-All Terrain Vehicle
MEDCAP	Medical Civic Assistance Programs
MEDEVAC	Medical Evacuation
MGS	Mobile Gun System
MICLIC	Mine Clearing Line Charge
MOLLE	Modular Lightweight Load-Carrying Equipment
MRAP	Mine Resistant Ambush Protected
MREs	Meals Ready to Eat
MTOE	Modified Table of Organization and Equipment
NAIs	Named Area of Interest
NCO	Non-Commissioned Officer
NGO	Nongovernmental Organizations
OEF	Operation Enduring Freedom
OHDACA	Overseas Humanitarian, Disaster, and Civic Aid
OP	Observation Post
PCI	Pre-Combat Inspection
PRTs	Provincial Reconstruction Teams
QRF	Quick Reaction Force
RC	Regional Command
RCP	Route Clearance Package
RPG	Rocket-Propelled Grenades
RTO	Radio Telephone Operator
SAW	Squad Automatic Weapon
SOF	Special Operations Forces
TAA	Tactical Assembly Area
TF	Task Force
TOC	Tactical Operations Center
UAV	Unmanned Aerial Vehicle
UNAMA	United Nations Assistance Mission in Afghanistan
USAID	United States Agency for International Development
WIA	Wounded in Action

About the Authors

Anthony E. Carlson holds a Ph.D. in History from the University of Oklahoma. He currently serves as an historian on the Afghan Study Team at the Combat Studies Institute and an adjunct Assistant Professor of History for the US Army Command and General Staff College. His publications include works on Progressive Era US water and flood control policy, public works, and the antebellum Army Corps of Topographical Engineers.

Michael J. Doidge is a Doctoral Candidate at the University of Southern Mississippi, where he co-edited *Triumph Revisited: Historians Battle for the Vietnam War*. In addition to writing his dissertation, Michael currently serves as an Army historian on the Afghan Study Team at the Combat Studies Institute and an adjunct Assistant Professor of History for the US Army Command and General Staff College.

Scott J. Gaitley holds a BA in History from Park University. He has served as an Air Force Wing Historian in both Iraq and Afghanistan and more recently as a staff historian at the Air Force Reserve Command Headquarters. He currently serves as an historian for the Afghan Study Team at the US Army Combat Studies Institute.

Kevin M. Hymel holds an MA in History from Villanova University. He is the author of Patton's *Photographs: War As He Saw It* and coauthor of *Patton: Legendary World* War II Commander, with Martin Blumenson. Before serving on the Afghan Study Team at the Combat Studies Institute, he worked for a number of military and military history magazines as a researcher, editor and writer.

Matt M. Matthews has worked at the Combat Studies Institute since July 2005 and is currently a member of the Afghan Study Team. He is the author of numerous CSI publications. Mr. Matthews has also coauthored various scholarly articles on the Civil War in the trans-Mississippi. He is a frequent speaker at Civil War Roundtables and the former Mayor of Ottawa, Kansas.

John M. McGrath holds an MA in History from the University of Massachusetts at Boston and is a Ph.D. candidate at Kansas State University. He is a retired Army officer and has worked as a US Army historian since 1998 at the Center of Military History and CSI. He is the author of nine books on military history.

Lieutenant Colonel John C. Mountcastle holds a Ph.D in History from Duke University. He has served as an Assistant Professor of History at the US Military Academy at West Point and more recently as an historian at the US Army Combat Studies Institute. He currently serves as the Professor of Military Science at the University of Washington in Seattle.

Ryan D. Wadle received a Ph.D. in History from Texas A&M University. He currently serves as an historian on the Afghan Study Team at the US Army Combat Studies Institute. In addition to his official duties, he is currently working on an article-length study of joint Army-Navy training and doctrine in the interwar period.